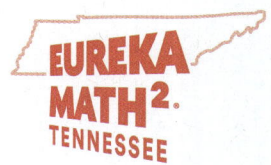

EUREKA MATH². TENNESSEE

A Story of Units®

Fractions Are Numbers ▸ 5

APPLY

Great Minds® is the creator of *Eureka Math*®,
Wit & Wisdom®, *Alexandria Plan*™, and *PhD Science*®.

Published by Great Minds PBC.
greatminds.org

Printed in the USA
B-Print

1 2 3 4 5 6 7 8 9 10 CCM 27 26 25 24 23

ISBN 978-1-63898-544-0

Contents

Place Value Concepts for Multiplication and Division with Whole Numbers

FAMILY MATH

Place Value Understanding for Whole Numbers

Dear Family,

Your student is learning to multiply and divide by 10, 100, and 1,000. They begin by representing multiplication and division on the place value chart. Students recognize patterns in products and quotients, which prepares them to calculate mentally. They write repeated multiplication by using exponents and explore how powers of 10 relate to place value and metric units. Your student solves problems by converting between metric measurements and describing the relationships between the units.

Key Terms

centigram	kiloliter
centiliter	milligram
convert	millimeter
exponent	mixed units
exponential form	power of 10

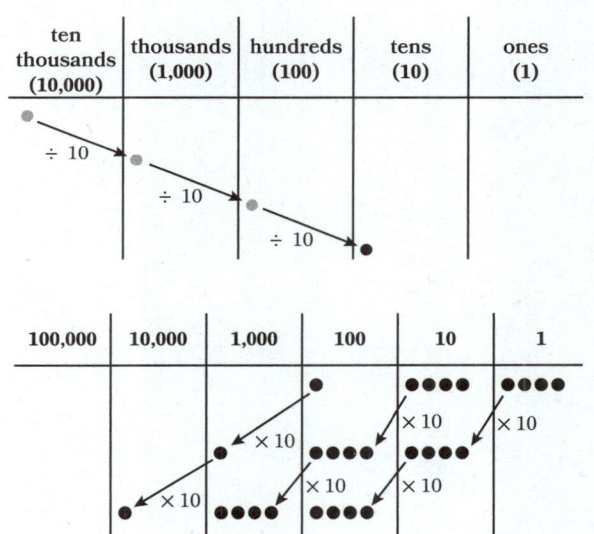

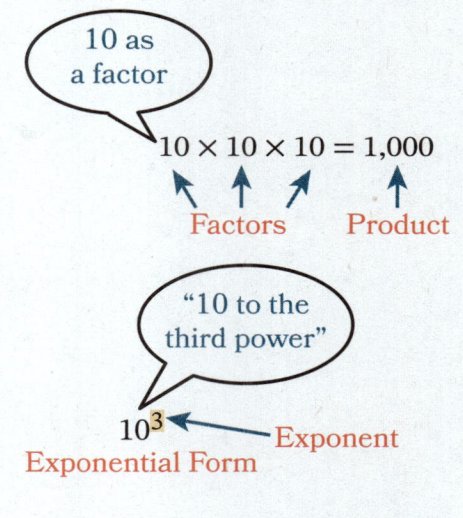

The place value chart shows each place value unit is 10 times as much as the unit to its right. This understanding can be applied to multiplying and dividing by powers of 10.

When multiplying by 10, digits shift to the left. When dividing by 10, digits shift to the right. The number of 10s can be used to find the product or the quotient just by shifting the digits.

kilometer, meter, centimeter, millimeter	kilogram, gram, centigram, milligram	kiloliter, liter, centiliter, milliliter
longest ⟶ shortest	heaviest ⟶ lightest	greatest capacity ⟶ least capacity
1 meter = 100 centimeters	1 gram = 100 centigrams	1 liter = 100 centiliters
1 meter = 1,000 millimeters	1 gram = 1,000 milligrams	1 liter = 1,000 milliliters
1 kilometer = 1,000 meters	1 kilogram = 1,000 grams	1 kiloliter = 1,000 liters

Students use powers of 10 to understand relationships between metric units and to solve problems.

At-Home Activity

Would You Rather?

Help your student practice converting metric units by asking "Would you rather" questions. For example, you could ask some of the following questions, replacing the pizza, chocolate milk, or scooter with your student's favorite items. As they answer each question, have them explain why.

- "Would you rather eat 100 grams of pizza or 10,000 centigrams of pizza?"

- "Would you rather drink 1 liter of chocolate milk or 1,000 milliliters of chocolate milk?"

- "Would you rather ride on a scooter for 5 kilometers or ride on a scooter for 500,000 centimeters?"

1

Name _____   Date _____

1. Use the place value chart to complete the statement and equation.

millions	hundred thousands	ten thousands	thousands	hundreds	tens	ones
		× 10	• • • •			

4 ten thousands is 10 times as much as __4 thousands__ .

$40,000 = 10 \times$ ____4,000____

Units on the place value chart follow a pattern. Beginning with the ones and moving left, each adjacent unit to the left is 10 times as much as the unit to the right.

The value of 4 in the ten thousands place is 10 times as much as the value of 4 in the thousands place.

2. Use the place value chart to complete the equation.

millions	hundred thousands	ten thousands	thousands	hundreds	tens	ones
		••••• ÷ 10	•••••			

$90,000 \div 10 =$ ____9,000____

Each adjacent unit to the right is 10 times as small as the unit to the left.

The value of 9 in the thousands place is 10 times as small as the value of 9 in the ten thousands place.

Use the place value chart to complete problems 3–5.

millions	hundred thousands	ten thousands	thousands	hundreds	tens	ones
6	5	5	7	8	4	7

3. The 6 in 6,557,847 represents ___6,000,000___ .

4. ___500,000___ ÷ 10 = 50,000

5. 7 thousands is ___1,000___ times as much as 7 ones.

> The 6 is in the millions place. The value of the 6 is 6,000,000.
>
> The value of 5 in the ten thousands place is 10 times as small as the value of 5 in the hundred thousands place.
>
> Thousands is 3 place value units to the left of ones. Each place value unit to the left is 10 times as much as the unit to the right. So 7 thousands is 10 × 10 × 10, or 1,000, times as much as 7 ones.

REMEMBER

Use the Read–Draw–Write process to solve the problem.

6. Box A has 72 paper clips in it. Box A has 4 times as many paper clips as box B. How many paper clips does box B have?

72 ÷ 4 = 18

Box B has 18 paper clips.

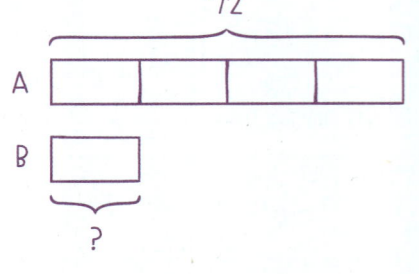

> I read the problem. I read again.
>
> As I reread, I think about what I can draw.
>
> I draw two tape diagrams: one to represent box A and one to represent box B.
>
> I draw the tape diagram for box A to show box A has 4 times as many paper clips as box B. I label the tape diagram for box A to show the total of 72 paper clips. I need to find the number of paper clips in box B, so I label it with a question mark.
>
> I divide to find the number of paper clips in box B.

© Great Minds PBC

1

Name _____ Date _____

1. Use the place value chart to complete the statement and equation.

millions	hundred thousands	ten thousands	thousands	hundreds	tens	ones

6 ten thousands is 10 times as much as _____ .

$60,000 = 10 \times$ _____

2. Use the place value chart to complete the equation.

millions	hundred thousands	ten thousands	thousands	hundreds	tens	ones

$50,000 \div 10 =$ _____

Use the place value chart to complete problems 3–5.

millions	hundred thousands	ten thousands	thousands	hundreds	tens	ones
9	8	8	3	4	7	3

3. The 9 in 9,883,473 represents _____ .

4. _____ $\div 10 = 80,000$

5. 3 thousands is _____ times as much as 3 ones.

REMEMBER

Use the Read–Draw–Write process to solve the problem.

6. Mrs. Chan has 96 bottles of paint. She has 8 times as many bottles of paint as paintbrushes. How many paintbrushes does Mrs. Chan have?

_____   _____
Name   Date

2

Multiply or divide.

1. a. $53 \times 10 =$ ___530___

    b. $53 \times 100 =$ ___5,300___

    c. $53 \times 1,000 =$ ___53,000___

2. a. $93,000 \div 10 =$ ___9,300___

    b. $93,000 \div 100 =$ ___930___

    c. $93,000 \div 1,000 =$ ___93___

3. $12 \times 100 =$ ___1,200___

4. $74,000 \div 10 =$ ___7,400___

5. $81 \times$ ___1,000___ $= 81,000$

6. $3,900 \div$ ___100___ $= 39$

> I notice patterns when multiplying by 10, 100, and 1,000. Each product is 10 times as much as the previous product.
>
> Multiplying by 100 is equivalent to multiplying by 10 twice.
>
> $$53 \times 10 \times 10 = 5,300$$
>
> Multiplying by 1,000 is equivalent to multiplying by 10 three times.
>
> $$53 \times 10 \times 10 \times 10 = 53,000$$

> There are similar patterns when dividing by 10, 100, and 1,000.
>
> Dividing by 100 is equivalent to dividing by 10 twice.
>
> $$93,000 \div 10 \div 10 = 930$$
>
> Dividing by 1,000 is equivalent to dividing by 10 three times.
>
> $$93,000 \div 10 \div 10 \div 10 = 93$$

> To find the product or quotient, I think about how many times to multiply or divide by 10.
>
> Each time I multiply by 10, the digits in the other factor shift to the left one time.
>
> Each time I divide by 10, the digits in the total shift to the right one time.
>
> To find the unknown product or divisor, I think about how many times the digits shift to the left or right.
>
> The digits in 81 shift to the left three times to make 81,000. This is the same as multiplying by 10 three times, or 1,000.
>
> The digits in 3,900 shift to the right two times. This is the same as dividing by 10 two times, or 100.

REMEMBER

7. The width of a rectangular patio is 8 feet. The area of the patio is 72 square feet. What is the length?

The length of the patio is 9 feet.

I know the width and the area of the patio. I need to find the length of the patio. I draw and label an area model to represent the patio.

8 ft

l ft

Area: 72 square feet

The formula for finding the area of a rectangle is $A = l \times w$.

$$72 = l \times 8$$

To find the length, I think about what number I multiply by 8 to get 72, or I can divide 72 by 8.

_____ _____

Name   Date

2

Multiply or divide.

1. a. $38 \times 10 =$ _____

    b. $38 \times 100 =$ _____

    c. $38 \times 1{,}000 =$ _____

2. a. $87{,}000 \div 10 =$ _____

    b. $87{,}000 \div 100 =$ _____

    c. $87{,}000 \div 1{,}000 =$ _____

3. $74 \times 10 =$ _____

4. $630{,}000 \div 100 =$ _____

5. $52 \times$ _____ $= 5{,}200$

6. $94{,}000 \div$ _____ $= 94$

REMEMBER

7. The width of a rectangular garden is 6 feet. The area of the garden is 48 square feet. What is the length?

Name _____ Date _____

Write each product or quotient in exponential form.

1. $10 \times 10 \times 10 \times 10 \times 10 \times 10 = $ ___ 10^6 ___

2. $100 \times 1,000 = $ ___ 10^5 ___

3. $1,000 \div 10 = $ ___ 10^2 ___

> A number that can be written as a product of 10s is a **power of 10**.
>
> I can also write a power of 10 in **exponential form**.
>
> The **exponent** represents how many times we use a number as a factor.
>
> There are six factors of 10, so the exponent is 6.

> I know $1,000 \div 10 = 100$.
>
> 100 in exponential form is 10^2.

> For $100 \times 1,000$, I rewrite each power of 10 as factors of 10.
>
> $100 = 10 \times 10$
>
> $1,000 = 10 \times 10 \times 10$
>
> $100 \times 1,000 = 10 \times 10 \times 10 \times 10 \times 10$
>
> I write this in exponential form as 10^5.

Find each product or quotient and write it in standard form.

4. $8 \times 10^4 = $ ___ $80,000$ ___

> I know multiplying 8 by 10^4 means shifting 8 to the left four times.

5. $700,000 \div 10^5 =$ ___7___

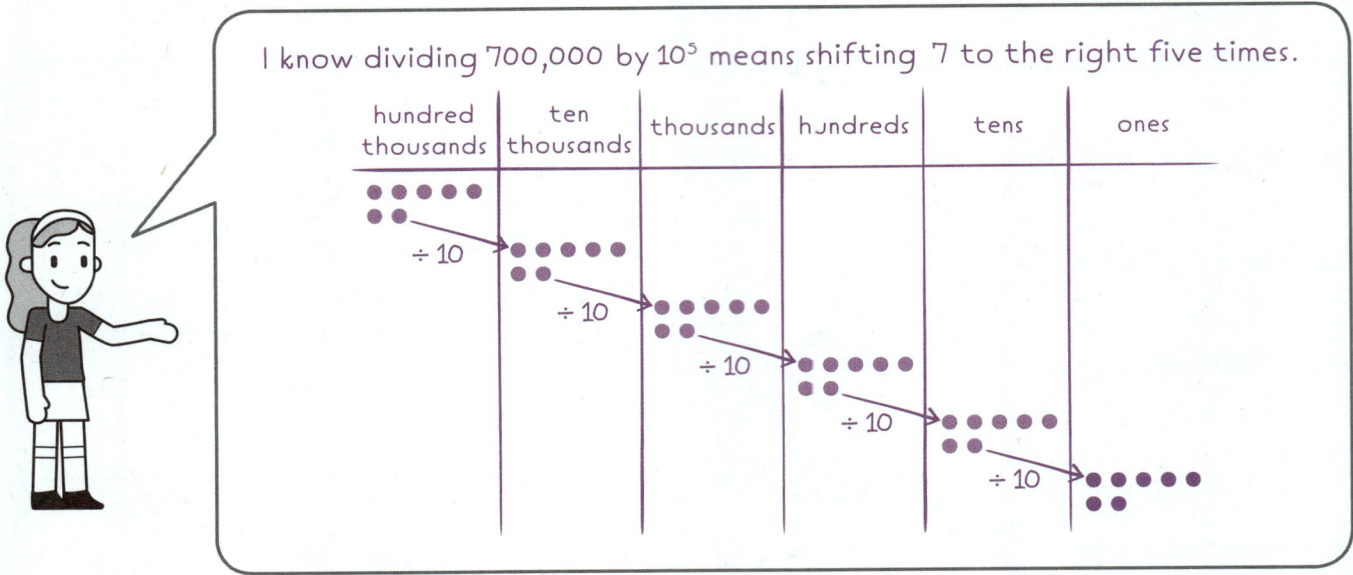

I know dividing 700,000 by 10^5 means shifting 7 to the right five times.

hundred thousands	ten thousands	thousands	hundreds	tens	ones

÷ 10

÷ 10

÷ 10

÷ 10

÷ 10

REMEMBER

6. Write an equation by using multiplication to show $\frac{3}{10} = \frac{30}{100}$.

$$\frac{3}{10} = \frac{10 \times 3}{10 \times 10} = \frac{30}{100}$$

I write an equation by using multiplication. I multiply 10 by the numerator and the denominator of $\frac{3}{10}$ to rename tenths as hundredths.

7. Write the decimal fraction in decimal form.

26 hundredths

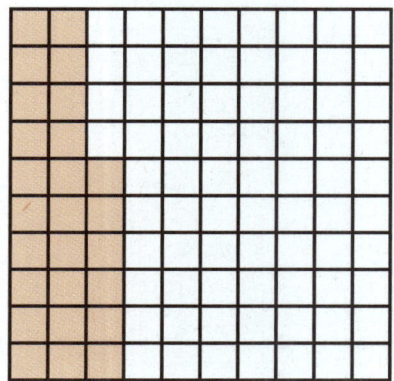

I can decompose 1 one into 100 hundredths. Each square represents 1 hundredth.

I know the model represents 26 hundredths.

I write 26 hundredths as a decimal number.

$\frac{26}{100} = \underline{\quad 0.26 \quad}$

Name _____   Date _____

Write each product or quotient in exponential form.

1. $10 \times 10 \times 10 \times 10 \times 10 =$ _____

2. $10 \times 100 =$ _____

3. $100{,}000 \div 10 =$ _____

Find each product or quotient and write it in standard form.

4. $7 \times 10^5 =$ _____

5. $60{,}000 \div 10^4 =$ _____

REMEMBER

6. Write an equation by using multiplication to show $\frac{8}{10} = \frac{80}{100}$.

7. Write the decimal fraction in decimal form.

64 hundredths

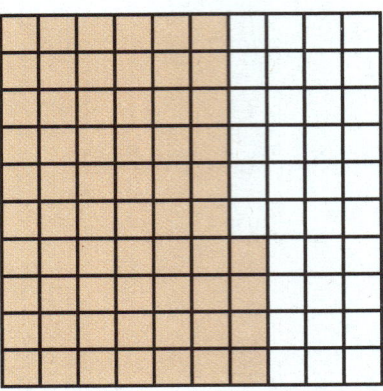

$\frac{64}{100} =$ _____

Name _____ **Date** _____

Estimate the product. Show your thinking.

1. $1{,}243 \times 33$

    $1{,}243 \times 33 \approx 1{,}200 \times 30$

    $= 36{,}000$

> I round the factors, and then use mental math to multiply.
>
> I analyze, or look carefully at, the factors to decide how to round.
>
> I round 1,243 to 1,200.
>
> I round 33 to 30.
>
> $$1{,}200 \times 30 = 12 \times 10^2 \times 3 \times 10$$
>
> I know $12 \times 3 = 36$.
>
> $10^2 \times 10 = 10^3$. I shift the digits in 36 three times to the left to multiply by 10^3.
>
> So $1{,}243 \times 33$ is about 36,000.

Estimate the quotient. Show your thinking.

2. $2{,}462 \div 5$

    $2{,}462 \div 5 \approx 2{,}500 \div 5$

    $= 500$

> I want to round 2,462 to a number divisible by 5.
>
> I round 2,462 to 2,500 because 25 hundreds is divisible by 5.
>
> 25 hundreds $\div$ 5 = 5 hundreds

3. The table shows the cost of tickets for a play.

Adult Ticket	Child Ticket
$28	$17

a. There are 5,321 adults at the play. About how much was spent on adult tickets?

$$5,321 \times 28 \approx 5,000 \times 30$$

$$= 150,000$$

About $150,000 was spent on adult tickets.

b. The total amount spent on child tickets was $7,616. About how many children are at the play?

$$7,616 \div 17 \approx 8,000 \div 20$$

$$= 400$$

About 400 children are at the play.

I want to use mental math. I round 5,321 to 5,000. I round 28 to 30. I multiply 5,000 and 30.

I think about the expression as 5 × 3 × 1,000 × 10.

5,321 × 28 is about 150,000.

To use mental math to divide, I round the numbers so the total is a multiple of the divisor.

I round $7,616 to $8,000. I round $17 to $20. I use mental math to find 8,000 ÷ 20.

I think about 8,000 ÷ 20 as 8,000 ÷ 10 ÷ 2.

7,616 ÷ 17 is about 400.

REMEMBER

4. What is the measure, in degrees, of an angle that is $\frac{43}{360}$ of a turn through a circle?

43°

One degree of an angle is equal to $\frac{1}{360}$ of a turn.

The denominator is 360 because there are 360 degrees in a circle.

The numerator tells me the measure of the angle in degrees.

4

Name _____ Date _____

Estimate each product. Show your thinking.

1. 368×11

2. $1{,}475 \times 52$

Estimate each quotient. Show your thinking.

3. $182 \div 3$

4. $3{,}155 \div 4$

5. The table shows the cost of tickets for a baseball game.

Adult Ticket	Child Ticket
$36	$28

a. There are 7,205 adults at the game. About how much was spent on adult tickets?

b. The total amount spent on child tickets was $6,328. About how many children are at the game?

REMEMBER

6. What is the measure, in degrees, of an angle that is $\frac{92}{360}$ of a turn through a circle?

　　　　　　　　© Great Minds PBC　•

Name _____   Date _____

1. Convert each measurement. Write an expression to help you convert. The first one is started for you.

Liters (L)	Expression	Centiliters (cL)
7	7×10^2	700
21	21×10^2	2,100

Convert means to express a measurement in terms of a different related measurement unit. **Kiloliters**, liters, **centiliters**, and milliliters are metric units for measuring capacity or liquid volume.

I know 1 liter = 100 centiliters.

I rename 1 L as 100 cL, or 10^2 cL, to convert between units.

$$7\ L = 7 \times 1\ L$$
$$= 7 \times 10^2\ cL$$
$$= 700\ cL$$

$$21\ L = 21 \times 1\ L$$
$$= 21 \times 10^2\ cL$$
$$= 2,100\ cL$$

Convert.

2. 1,400 cg = _____14,000_____ mg

Kilograms, grams, **centigrams**, and **milligrams** are metric units for measuring weight.

I know 1 centigram = 10 milligrams.

I rename 1 cg as 10 mg to convert between units.

$$1{,}400 \text{ cg} = 1{,}400 \times 1 \text{ cg}$$
$$= 1{,}400 \times 10 \text{ mg}$$
$$= 14{,}000 \text{ mg}$$

3. 700 m = _____700,000_____ mm

Kilometers, meters, centimeters, and **millimeters** are metric units for measuring length.

I know 1 meter = 1,000 millimeters.

I rename 1 m as 1,000 mm, or 10^3 mm, to convert between units.

$$700 \text{ m} = 700 \times 1 \text{ m}$$
$$= 700 \times 10^3 \text{ mm}$$
$$= 700{,}000 \text{ mm}$$

REMEMBER

4. Use a protractor to draw an angle that has a measure of 52°.

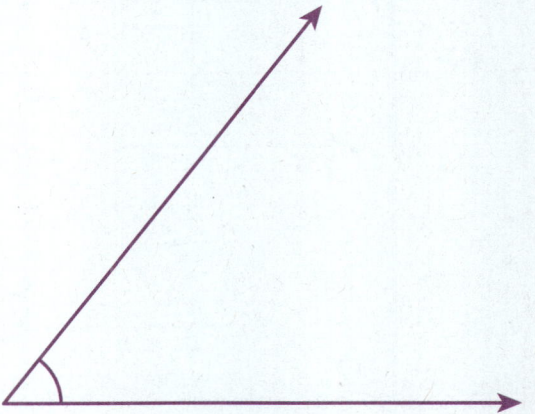

I use my straightedge to draw a ray.

I line up my protractor with the ray so the endpoint is where the zero line and the 90° line meet and the ray crosses through the 0° tick mark.

I find the tick mark that represents 52° on the scale of the protractor. Then I draw a tick mark on my paper that lines up with the 52° tick mark.

I use my straightedge to draw the second ray from the end point of the first ray to the tick mark. I draw an arrowhead at the end of the second ray.

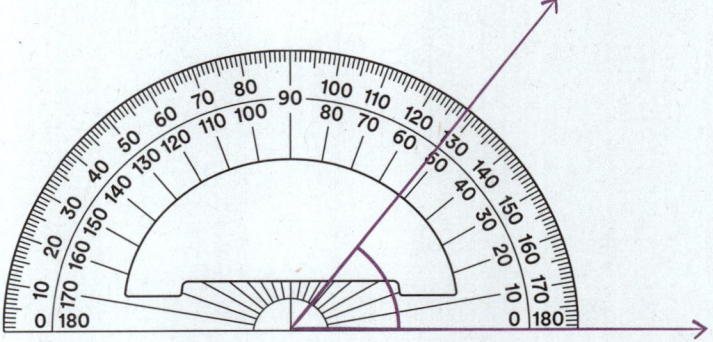

I draw an arc to show the angle I drew. The angle is acute, so the arc shows a measure less than 90°.

Name _____ Date _____

Convert each measurement. Write an expression to help you convert. The first one is started for you.

1.

Meters (m)	Expression	Millimeters (mm)
6	6×10	_____
12	_____	_____
305	_____	_____
540	_____	_____

2.

Liters (L)	Expression	Centiliters (cL)
8	8×10	_____
31	_____	_____
450	_____	_____
600	_____	_____

Convert.

3. 2,300 cg = _____ mg

4. 500 m = _____ cm

5. 70 kL = _____ L

REMEMBER

6. Use a protractor to draw an angle that has a measure of 37°. A paper protractor is included, if needed.

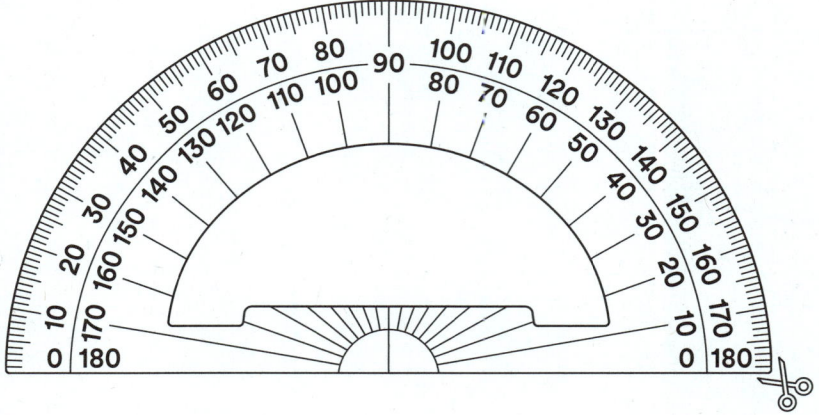

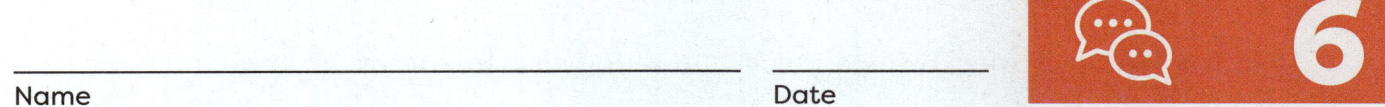

Name _____   Date _____

Use the Read–Draw–Write process to solve the problem.

1. Mr. Sharma pours fuel into 5 tanks. He pours 450 milliliters of fuel into each tank.

    a. About how many milliliters of fuel are in the tanks altogether?

    $5 \times 450 \approx 5 \times 500$

    $\qquad = 2,500$

    There are about 2,500 milliliters of fuel in the tanks.

    b. Exactly how many milliliters of fuel are in the tanks altogether?

    $5 \times 450 = 2,250$

    There are 2,250 milliliters of fuel in the tanks.

    c. How do you know your answer in part (b) is reasonable?

    My answer is reasonable because 2,250 milliliters is close to my estimate of 2,500 milliliters.

I read the problem. I read again.

As I reread, I think about what I can draw.

I draw a tape diagram to show 5 tanks as 5 equal parts. I label each part 450 to show Mr. Sharma pours 450 milliliters of fuel into each tank. I do not know the total, so I label the total with a question mark.

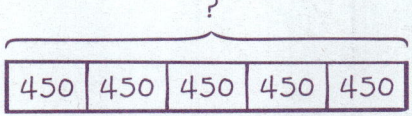

I estimate the total number of milliliters. 450 mL is about 500 mL, so I multiply to find 5 groups of 500.

To find the exact answer, I multiply to find 5 groups of 450.

I compare my exact answer to my estimate to see whether it is reasonable.

Use the Read–Draw–Write process to solve the problem.

2. Toby has 426 cm of twine. He uses 1 m 56 mm of the twine. How many millimeters of twine does Toby have left?

1 m 56 mm = 1,056 mm

426 cm = 4,260 mm

4,260 mm − 1,056 mm = 3,204 mm

Toby has 3,204 millimeters of twine left.

I read the problem. I read again.

As I reread, I think about what I can draw.

I draw a tape diagram to represent how much twine Toby has and how much he uses. I label the part that represents how many millimeters of twine Toby has left with a question mark.

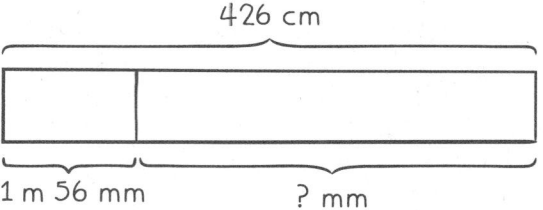

426 cm

1 m 56 mm   ? mm

The lengths are not written in the same unit, so I need to convert the measurements.

I convert 426 cm to millimeters.

$$426 \text{ cm} = 426 \times 1 \text{ cm}$$
$$= 426 \times 10 \text{ mm}$$
$$= 4,260 \text{ mm}$$

I convert 1 m 56 mm to millimeters.

1 m = 1,000 mm

1,000 mm + 56 mm = 1,056 mm

Then I subtract the 1,056 mm of twine Toby uses from the 4,260 mm he started with to find how much he has left.

REMEMBER

3. Use $\overleftrightarrow{MN}$ to complete parts (a) and (b).

Sample:

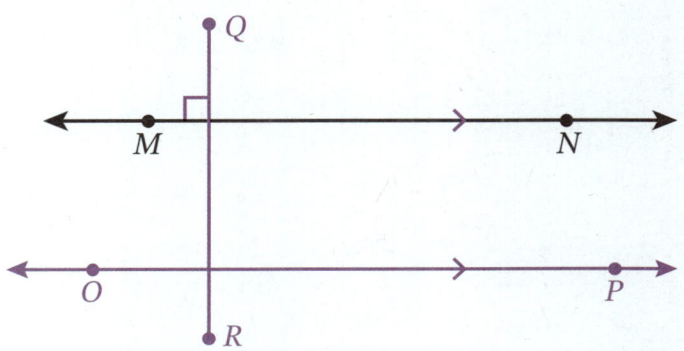

a. Draw $\overleftrightarrow{OP}$ parallel to $\overleftrightarrow{MN}$.

b. Draw $\overline{QR}$ perpendicular to $\overleftrightarrow{MN}$.

I use a right-angle tool and a straightedge to draw a parallel line. Parallel lines never intersect.

I line up my right-angle tool along $\overleftrightarrow{MN}$ so the top edge touches the line. Then I hold my straightedge along the other edge of the right-angle tool.

I slide my right-angle tool down along the straightedge. I trace along the top edge to draw line $\overleftrightarrow{OP}$. I mark the lines to show they are parallel.

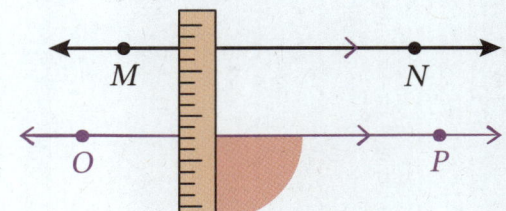

$\overleftrightarrow{OP}$ extends beyond point O and point P. I use arrowheads on both ends to identify it as a line.

I use a right-angle tool and a straightedge to draw a perpendicular line segment. To be perpendicular, the line and the line segment intersect at right angles.

I line up one edge of the right-angle tool along $\overleftrightarrow{MN}$.

I trace along the edge of the right-angle tool that is perpendicular to $\overleftrightarrow{MN}$.

I label the ends of the line segment with point Q and point R. I mark the angle to show the lines are perpendicular.

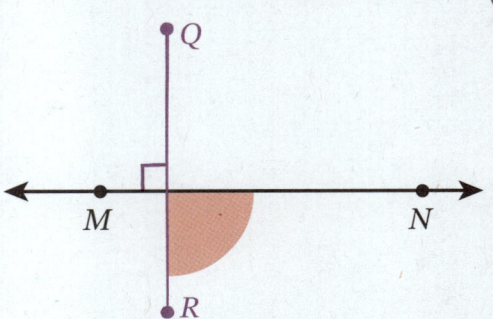

6

Name _____ Date _____

Use the Read–Draw–Write process to solve each problem.

1. Miss Baker pours lemonade into 6 pitchers. She pours 350 milliliters of lemonade into each pitcher.

   a. About how many milliliters of lemonade are in the pitchers altogether?

   b. Exactly how many milliliters of lemonade are in the pitchers altogether?

   c. How do you know your answer in part (b) is reasonable?

2. Sasha has 520 cm of rope. She uses 2 m 72 mm of the rope. How many millimeters of rope does Sasha have left?

REMEMBER

3. Use $\overleftrightarrow{AB}$ to complete parts (a) and (b).

a. Draw $\overline{CD}$ parallel to $\overleftrightarrow{AB}$.

b. Draw $\overleftrightarrow{EF}$ perpendicular to $\overleftrightarrow{AB}$.

FAMILY MATH
Multiplication of Whole Numbers

Dear Family,

Your student is learning to multiply whole numbers more efficiently. They use models and methods from earlier grades such as area models, break apart and distribute, partial products, and the standard algorithm. Your student multiplies multi-digit numbers and makes connections between different strategies. These connections support their work using the standard algorithm to multiply larger numbers.

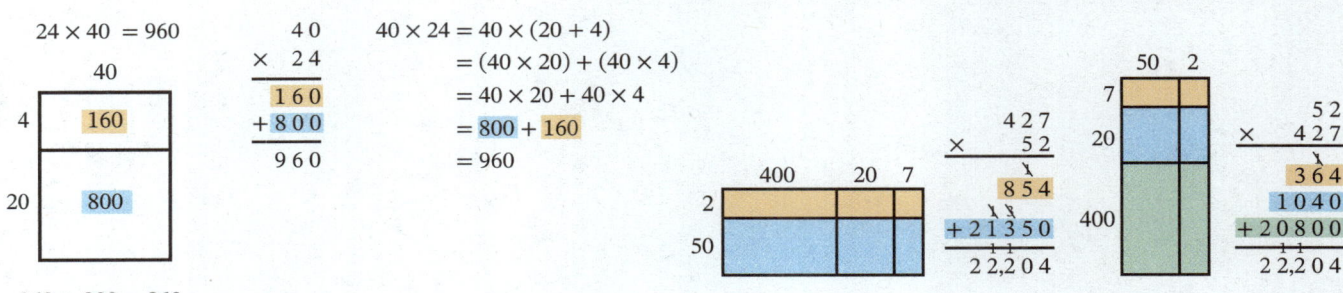

Different methods can be used to solve the same multiplication problem. In each method factors are decomposed, each part is multiplied, and the partial products are added to find the total.

The number of rows in the area model is the same as the number of partial products in the standard algorithm.

At-Home Activity

How Many Hours Away?

Help your student use multiplication to calculate the number of hours until a fun event. Ask your student to think of an event they are looking forward to such as their birthday, a holiday, a vacation, or a family gathering. Encourage them to choose something that is at least 1 month away. Figure out how many days there are until the event and multiply that number by the number of hours in a day, as in the following example.

- There are 43 days until a family reunion.

- Multiply the number of days until the event by the number of hours in a day:
  $43 \times 24 = 1,032$.

- Next, add on any remaining hours left today to find the total number of hours until the event. For example, if it is 5:00 p.m., there are 7 more hours left in the day. Find $1,032 + 7$ to determine there are $1,039$ hours until the family reunion.

For an additional challenge, have your student find the number of minutes until the event.

- Multiply the number of hours until the event by the number of minutes in an hour:
  $1,039 \times 60 = 62,340$. There are $62,340$ minutes until the family reunion.

7

Name _____ Date _____

Multiply. Show or explain your strategy.

1. 2 times as much as 741

	700	40	1
2	1,400	80	2

$2 \times 741 = 1{,}400 + 80 + 2 = 1{,}482$

> 2 times as much as 741 is equivalent to 2 × 741.
>
> I draw an area model to represent the problem.
>
> I label the width of the area model as 2.
>
> I decompose the length of the area model, 741, into three parts: 700, 40, and 1.
>
> | | 700 | 40 | 1 |
> |---|---|---|---|
> | 2 | | | |
>
> I multiply each part by 2.
>
> $2 \times 700 = 1{,}400$
>
> $2 \times 40 = 80$
>
> $2 \times 1 = 2$
>
> | | 700 | 40 | 1 |
> |---|---|---|---|
> | 2 | 1,400 | 80 | 2 |
>
> I add the partial products.

2. 6 times as long as 3,058 kilometers

$$6 \times 3,058 = 6 \times (3,000 + 50 + 8)$$
$$= 6 \times 3,000 + 6 \times 50 + 6 \times 8$$
$$= 18,000 + 300 + 48$$
$$= 18,348$$

18,348 kilometers

6 times as long as 3,058 is equivalent to $6 \times 3,058$.

I decompose 3,058 by place value units.

$$3,058 = 3,000 + 50 + 8$$

I multiply each part by 6.

I add the partial products together.

3. $9 \times 54,967$

```
   5 4,9 6 7
 ×         9
   4 8 6 6
 4 9 4,7 0 3
```

I write $9 \times 54,967$ vertically.

Starting with the ones and moving left, I multiply each place value unit of 54,967 by 9.

```
   5 4,9 6 7
 ×         9
         6
           3
```

9 ones × 7 ones = 63 ones, so I rename 63 ones as 6 tens 3 ones, and then record 3 in the ones place of the product. I record 6 in the tens column above the product.

9 ones × 6 tens = 54 tens, so I add 54 tens and 6 tens, and then record 60 tens as 6 hundreds 0 tens.

```
   5 4,9 6 7
 ×         9
       6 6
         0 3
```

I continue multiplying each place value unit by 9 and adding units as I go to get the final product.

REMEMBER

4. $\angle LNQ$ is a straight angle. Write and solve an equation to find the unknown angle measure.

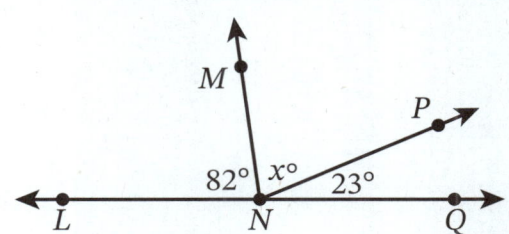

$$82 + x + 23 = 180$$
$$x + 105 = 180$$
$$x = 75$$

The measure of $\angle MNP$ is __75°__ .

A straight angle has a measure of 180°. So the sum of the angle measures is 180°.

I add the known angle measures.

$$82 + 23 = 105$$

Then I subtract that from the total angle measure, 180°, to find the unknown angle measure.

$$180 - 105 = 75$$

5. Use >, =, or < to compare the decimal numbers. Plot the location of each decimal number on the number line to justify your answer.

19.67 ___>___ 19.47

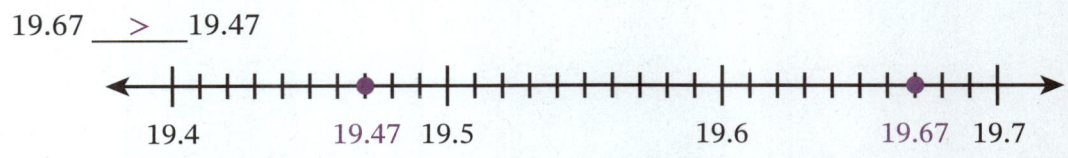

I see tenths labeled on the number line. Each 1 tenth is decomposed into 10 hundredths. So the distance between each tick mark on the number line represents 1 hundredth.

I know 19.67 has 6 tenths 7 hundredths. I start at 19.6 and count on 7 hundredths. I draw a dot on the number line and label the tick mark 19.67.

I know 19.47 has 4 tenths 7 hundredths. I start at 19.4 and count on 7 hundredths. I draw a dot on the number line and label the tick mark 19.47.

I see 19.67 is to the right of 19.47. I know 19.67 is greater than 19.47.

I write the symbol for greater than: >.

7

Name _____ Date _____

Multiply. Show or explain your strategy.

1. 3 times as much as 536

2. 8 times as long as 2,403 meters

3. $5 \times 16{,}521$

4. $7 \times 28{,}953$

REMEMBER

5. $\angle ACE$ is a straight angle. Write and solve an equation to find the unknown angle measure.

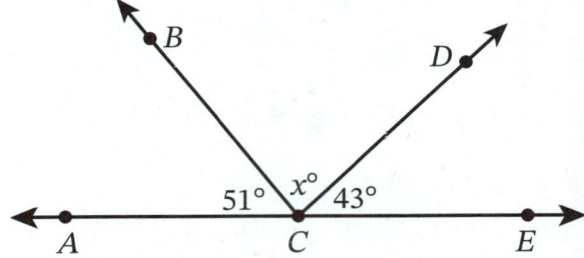

The measure of $\angle BCD$ is _____ .

---

6. Use >, =, or < to compare the decimal numbers. Plot the location of each decimal number on the number line to justify your answer.

11.24 _____ 11.04

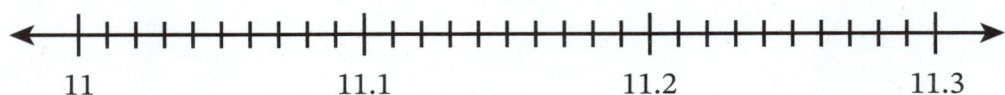

8

Name _____ Date _____

Complete the area model. Then multiply by showing two partial products.

1. 24×12

a.

```
          12
      ┌─────────┐
   4  │    48   │
      ├─────────┤
      │         │
  20  │   240   │
      │         │
      └─────────┘
```

b.

		1	2
×		2	4
		4	8
+	2	4	0
	2	8	8

I find the partial products in the area model by multiplying the length by each width: $4 \times 12 = 48$ and $20 \times 12 = 240$.

In vertical form, the unit is 12, so that factor is recorded first.

I write the partial products from the area model in vertical form.

To find the product, I add the partial products.

$48 + 240 = 288$, so $24 \times 12 = 288$.

2. 43×212

a.

212

| 3 | 636 |
| 40 | 8,480 |

b.

		2	1	2
×			4	3
		6	3	6
+	8	4	8	0
	9,	1	1	6

I make the unit 212, so I label the length of the rectangle 212.
I decompose 43 into 40 and 3. I label widths of the rectangles.

212

| 3 | |
| 40 | |

I find the partial products in the area model by multiplying the length by each width: $3 \times 212 = 636$ and $40 \times 212 = 8,480$.

212

| 3 | 636 |
| 40 | 8,480 |

In vertical form, the unit is 212, so that factor is recorded first.

I write the partial products from the area model in vertical form.

To find the product, I add the partial products.

$636 + 8,480 = 9,116$, so $43 \times 212 = 9,116$.

REMEMBER

3. Use the associative property to find factors.

$$42 = 7 \times \underline{\quad 6 \quad}$$

$$= 7 \times (2 \times \underline{\quad 3 \quad})$$

$$= (\underline{\quad 7 \quad} \times 2) \times 3$$

$$= \underline{\quad 14 \quad} \times 3$$

$$= \underline{\quad 42 \quad}$$

What are some factors of 42?

1, 2, 3, 6, 7, 14, 42

> I know two factors of 42 are 1 and the number itself, so 1 and 42 are factors.
>
> I know 42 = 7 × 6, so 7 and 6 are factors.
>
> I use the associative property to help me find more factors.
>
> I start with 42 = 7 × 6.   42 = 7 × 6
>
> I rename as 6 as 2 × 3.   = 7 × (2 × 3)
>
> I change the grouping of the factors.   = (7 × 2) × 3
>
> I multiply 2 by 7.   = 14 × 3
>
> I find 42 = 14 × 3, so 3 and 14 are factors.
>
> I also list 2 as a factor because 42 = 7 × 2 × 3.

8

Name _____

Date _____

Complete the area model. Then multiply by showing two partial products.

1. 58×20

   a.

   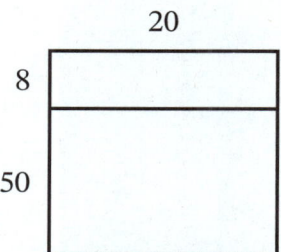

   b.

			2	0
×			5	8
+				

2. 22×34

   a.

   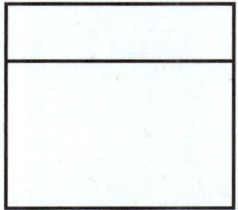

   b.

		3	4
×		2	2
+			

3. 32×133

   a.

   b.

		1	3	3
×			3	2
+				

REMEMBER

4. Use the associative property to find factors.

$60 = 6 \times$ _____

$ = (3 \times$ _____$) \times (2 \times$ _____$)$

$ = (2 \times 2) \times (3 \times$ _____$)$

$ = 4 \times$ _____

$ =$ _____

What are some factors of 60?

9

_____   _____
Name   Date

Complete the area model. Then multiply by using the standard algorithm.

1. 82 × 543

a.

	500	40	3
2	1,000	80	6
80	40,000	3,200	240

b.

			5	4	3
×				8	2
		1	0	8	6
+	4	³3̸	⁸4̸	4	0
	4	4,	5	2	6

I multiply the length and width of each rectangle in the area model to find the partial products. Then I add the partial products to find the product.

1,000 + 80 + 6 = 1,086 and 40,000 + 3,200 + 240 = 43,440.

1,086 + 43,440 = 44,526

I multiply 543 by 2 to find the first partial product in the standard algorithm.

I think in unit form to record the digits in the correct places.

2 ones × 3 ones = 6 ones   2 ones × 4 tens = 8 tens   2 ones × 5 hundreds = 10 hundreds

The first partial product is the same as the sum of the first row of the area model: 1,086.

I multiply 543 by 80 to find the second partial product.

8 tens × 3 ones = 24 tens. I record 24 tens as 2 hundreds 4 tens.

8 tens × 4 tens = 32 hundreds. I add 32 hundreds to the 2 hundreds already recorded and rename 34 hundreds as 3 thousands 4 hundreds.

8 tens × 5 hundreds = 40 thousands. I add the 3 thousands to get 43 thousands. I record it as 4 ten thousands 3 thousands.

The second partial product is the same as the sum of the second row of the area model: 43,440.

I add the two partial products in the algorithm. This gives me the same product from the area model.

Draw an area model to find the partial products. Then multiply by using the standard algorithm.

2. 67 × 631

a.

	600	30	1
7	4,200	210	7
60	36,000	1,800	60

b.

			6	3	1
×				6	7
		4	4	1	7
+	3	7	8	6	0
	4	2,	2	7	7

I decompose each factor by place value units.

631 = 600 + 30 + 1 and 67 = 60 + 7.

I want to think of the problem as 67 groups of 631 so there are only two partial products. I draw an area model with two rows of three rectangles.

	600	30	1
7			
60			

I multiply the length and width of each rectangle in the area model to find the partial products. To find the product, I add the partial products from my area model.

I multiply 631 by 7 to find the first partial product in the standard algorithm. I think about unit form to help me add and record units as I multiply.

The first partial product is the same as the sum of the first row of the area model: 4,417.

I multiply 631 by 60 to find the second partial product. I think about unit form to help me add and record units as I multiply.

The second partial product is the same as the sum of the second row of the area model: 37,860.

I add the two partial products in the algorithm. This gives me the same product from the area model.

REMEMBER

3. Think about the multiples of 3.

    a. Write the first 10 multiples of 3. Start with 3.

        __3__ , __6__ , __9__ , __12__ , __15__ , __18__ , __21__ , __24__ , __27__ , __30__

    b. What is the eighth multiple of 3?

        __24__

    c. Is 28 a multiple of 3?

        No.

> I skip-count by threes to find each multiple.
>
> I see the eighth multiple of 3 is 24. I can also find 8 × 3 to check that the eighth multiple is 24.
>
> I know 28 is not a multiple of 3 because I do not say 28 when I count by threes. I cannot multiply another whole number by 3 to get 28.

4. On Monday, Ray reads 4 pages of a book. Each day, he reads 4 more pages than the day before.

    a. Complete the table.

Day	Monday	Tuesday	Wednesday	Thursday	Friday
Number of Pages	4	8	12	16	20

    b. What patterns do you notice in the number of pages?

    Sample:

    The number of pages increases by 4 each day.

    Every other number in the pattern is a multiple of 8.

> I skip-count by fours to complete the table.
>
>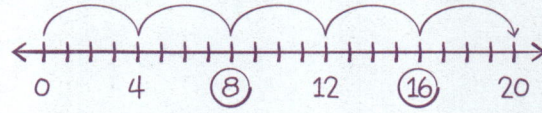
>
> So the number of pages increases by 4 each day.
>
> I look for another pattern.
>
> I know 4 is a factor of 8 and 2 × 4 = 8. So every other multiple of 4 is a multiple of 8.

Name _____   Date _____

Complete the area model. Then multiply by using the standard algorithm.

1. 48 × 36

   a.

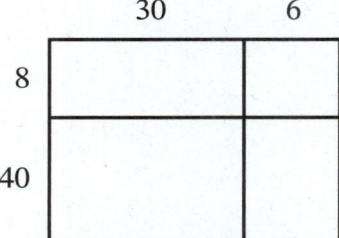

   b.

				3	6
	×			4	8
+					

2. 34 × 461

   a.

	400	60	1
4			
30			

   b.

			4	6	1
	×			3	4
+					

Draw an area model to find the partial products. Then multiply by using the standard algorithm.

3. 71 × 342

   a.

   b.

			3	4	2
	×			7	1
+					

4. Think about the multiples of 4.

   a. Write the first 10 multiples of 4. Start with 4.

   _____ , _____ , _____ , _____ , _____ , _____ , _____ , _____ , _____ , _____

   b. What is the seventh multiple of 4?

   _____

   c. Is 32 a multiple of 4?

5. On Monday, Jayla jumps rope for 3 minutes. Each day, she jumps rope for 2 more minutes than the day before.

   a. Complete the table.

Day	Monday	Tuesday	Wednesday	Thursday	Friday
Number of Minutes	3				

   b. What patterns do you notice in the number of minutes?

Name _____ **Date** _____

💬 **10**

Draw an area model to find the partial products and find their sum. Then multiply by using the standard algorithm. Compare your answers in each part to check that the product is correct.

1. 412 × 361

a.

	300	60	1
2	600	120	2
10	3,000	600	10
400	120,000	24,000	400

```
      7 2 2
    3,6 1 0
+ 1 4 4,4 0 0
  ‾‾‾‾1‾‾‾‾‾
  1 4 8,7 3 2
```

b.

					3	6	1
×					4	1	2
					7̷	2	2
				3	6	1	0
+	1	4̷	4	4	0	0	
	1	4	8,	7	3	2	

I decompose each factor in expanded form. 412: 400 + 10 + 2 and I label the parts along the left side of the area model. 361: 300 + 60 + 1 and I label the parts along the top of the area model. I find the areas of the nine rectangles by multiplying the length of each rectangle by its width.

I add the partial products in the area model.

	300	60	1	
2	600	120	2	→ 722
10	3,000	600	10	→ 3,610
400	120,000	24,000	400	→ 144,400

```
      7 2 2
    3, 6 1 0
+ 1 4 4, 4 0 0
  ‾‾‾‾1‾‾‾‾‾‾
  1 4 8, 7 3 2
```

In the standard algorithm, first I find 361 × 2. The partial product has the same value as the sum of the areas in the first row of the area model: 722.

Next, I find 361 × 10. The partial product has the same value as the sum of the areas in the second row of the area model: 3,610.

Then, I find 361 × 400. The partial product has the same value as the sum of the areas in the third row of the area model: 144,400.

Last, I add the three partial products.

Estimate the product. Then multiply.

2. $272 \times 1,205$

$272 \times 1,205 \approx$ <u> 300 </u> $\times$ <u> 1,000 </u>

$= $ <u> 300,000 </u>

$$
\begin{array}{r}
1,2\,0\,5 \\
\times\qquad 2\,7\,2 \\
\hline
2\,4\,1\,0 \\
8\,4\,3\,5\,0 \\
+\,2\,4\,1\,0\,0\,0 \\
\hline
3\,2\,7,7\,6\,0
\end{array}
$$

First, I estimate the product.

I round 272 to 300. I round 1,205 to 1,000.

Next, I multiply 272 and 1,205 by finding partial products.

I multiply each digit in 1,205 by 2. I hold on to place value in my head as I multiply.

I know 2 × 5 represents 2 ones × 5 ones, which equals 10 ones or 1 ten. I record 1 in the tens place and 0 in the ones place of the partial product.

I know 2 × 2 represents 2 ones × 2 hundreds, which equals 4 hundreds. I record 4 in the hundreds place of the partial product.

I know 2 × 1 represents 2 ones × 1 thousand, which equals 2 thousands. I record 2 in the thousands place of the partial product.

The first partial product is 2,410.

I hold on to place value in my head as I multiply each digit in 1,205 by 7. I know 7 represents 7 tens. I record the partial product 84,350.

I hold on to place value in my head as I multiply each digit in 1,205 by 200. I know 2 represents 2 hundreds. I record the partial product 241,000.

Last, I add the three partial products.

$$
\begin{array}{r}
2,4\,1\,0 \\
8\,4,3\,5\,0 \\
+\,2\,4\,1,0\,0\,0 \\
\hline
3\,2\,7,7\,6\,0
\end{array}
$$

The product 327,760 is close to my estimate of 300,000.

REMEMBER

3. Record the factor pairs for the given number as multiplication expressions. List the factors in order from least to greatest. Then circle prime or composite for the number.

Number	Multiplication Expressions	Factors	Prime or Composite
13	1×13	1, 13	(Prime) Composite

A whole number greater than 1 is a prime number if its only factors are 1 and itself.

A whole number greater than 1 is a composite number if it has more than two factors.

I make rectangular arrays of 13 circles to find the factor pairs.

One array has 1 row of 13 circles. This shows $1 \times 13 = 13$.

There are no other rectangular arrays that can be made with 13.

The only factor pair for 13 is 1×13.

Because the only factors are 1 and 13, the number is prime.

✎ **10**

Name _____ Date _____

Draw an area model to find the partial products and find their sum. Then multiply by using the standard algorithm. Compare your answers in each part to check that the product is correct.

1. 523×416

    a.

    b.

				4	1	6
×				5	2	3
+						

Estimate the product. Then multiply.

2. $244 \times 2{,}031$

    $244 \times 2{,}031 \approx$ _____ $\times$ _____

    $=$ _____

    $$\begin{array}{r} 2{,}0\ 3\ 1 \\ \times \quad\ 2\ 4\ 4 \\ \hline \end{array}$$

    $$\begin{array}{r} + \quad\quad\quad \\ \hline \end{array}$$

REMEMBER

3. Record the factor pairs for the given number as multiplication expressions. List the factors in order from least to greatest. Then circle prime or composite for the number.

Number	Multiplication Expressions	Factors	Prime or Composite
17			Prime Composite

11

Name _____   Date _____

Estimate the product. Then multiply.

1. $528 \times 749 \approx$ __500__ $\times$ __700__

    $=$ __350,000__

```
        5 2 8
   ×    7 4 9
        4̶ 7̶ 5 2
      2 1̶ 1̶ 2 0
   + 3̶ 6̶ 9 6 0 0
      1 1
      3 9 5, 4 7 2
```

To estimate the product, I round both factors to the nearest hundred and multiply.

Then I multiply to find the exact product. When I multiply to find each partial product, I can think about the factors in unit form, in standard form, or as single-digit by single-digit multiplication.

I find 9 × 528. The partial product is 4,752.

I find 40 × 528. The partial product is 21,120.

I find 700 × 528. The partial product is 369,600.

Last, I add the three partial products together.

```
      4, 7 5 2
      2 1, 1 2 0
   + 3 6 9, 6 0 0
      1 1
      3 9 5, 4 7 2
```

My estimated product is 350,000. The actual product is 395,472. This is reasonable because my estimated factors are both less than the actual factors.

Multiply.

2. $6,556 \times 762$

```
        6,5 5 6
×         7 6 2
    ̶1̶ ̶1̶ ̶1̶
    1 3 1 1 2
    ̶3̶ ̶3̶ ̶3̶
    3 9 3 3 6 0
    ̶3̶ ̶3̶ ̶4̶
+ 4 5 8 9 2 0 0
    ̶1̶ ̶1̶
    4,9 9 5,6 7 2
```

When I multiply to find each partial product, I can think about the factors in unit form, in standard form, or as single-digit by single-digit multiplication.

First, I find $2 \times 6,556$. The partial product is 13,112.

Next, I find $60 \times 6,556$. The partial product is 393,360.

Then, I find $700 \times 6,556$. The partial product is 4,589,200.

Last, I add the three partial products together.

```
      1 3,1 1 2
      3 9 3,3 6 0
  + 4,5 8 9,2 0 0
      ̶1̶ ̶1̶
      4,9 9 5,6 7 2
```

REMEMBER

3. Show that the pair of fractions are equivalent by drawing an area model. Then express the equivalence by using multiplication.

$\frac{2}{5}$ and $\frac{4}{10}$

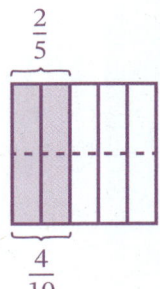

$\frac{2}{5} = \frac{2 \times 2}{2 \times 5} = \frac{4}{10}$

> The area model shows each fifth partitioned into 2 times as many parts.
>
> I multiply to show 2 times as many shaded parts and 2 times as many total parts.

> For $\frac{2}{5}$, I draw an area model partitioned into fifths and shade 2 equal parts.
>
> To show $\frac{4}{10}$, I draw a horizontal line across the middle of the area model.
>
> Now the area model is partitioned into 10 equal parts with 4 parts shaded.
>
> I see that $\frac{2}{5}$ is equivalent to $\frac{4}{10}$.

4. Complete the equation to show an equivalent fraction.

$$\frac{3}{4} = \frac{6}{8}$$

> I use a multiple of the denominator to find an equivalent fraction.
>
> I know 8 is a multiple of 4 and 4 × 2 = 8.
>
> I multiply the numerator and the denominator of $\frac{3}{4}$ by 2 to create an equivalent number of eighths.
>
> $\frac{3}{4} = \frac{2 \times 3}{2 \times 4} = \frac{6}{8}$

11

Name _____   Date _____

Estimate the product. Then multiply.

1. $828 \times 237 \approx$ _____ × _____

    = _____

$$\begin{array}{r} 8\ 2\ 8 \\ \times\quad 2\ 3\ 7 \\ \hline \end{array}$$

2. $574 \times 417 \approx$ _____ × _____

    = _____

$$\begin{array}{r} 5\ 7\ 4 \\ \times\quad 4\ 1\ 7 \\ \hline \end{array}$$

Multiply.

3. $6{,}281 \times 286$

4. $4{,}898 \times 913$

REMEMBER

5. Show that the pair of fractions are equivalent by drawing an area model. Then express the equivalence by using multiplication.

$\frac{1}{5}$ and $\frac{2}{10}$

$$\frac{1}{5} =$$

6. Complete the equation to show an equivalent fraction.

$$\frac{5}{6} = \frac{}{12}$$

FAMILY MATH
Division of Whole Numbers

Dear Family,

Your student is learning to find quotients of multi-digit dividends and divisors. Earlier in grade 4, they used a variety of methods to solve division problems with one-digit divisors. Now in grade 5, your student uses these familiar methods to divide by a two-digit divisor. They use area models to help them solve and understand the vertical form of division. They also use familiar estimation strategies that help them determine whether their answers make sense. Students solve word problems and explain the meaning of quotients and remainders, which is important later in grade 5 when they solve multi-step word problems.

Key Term

dividend

$$926 \div 23$$

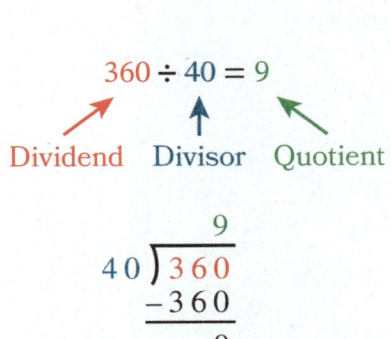

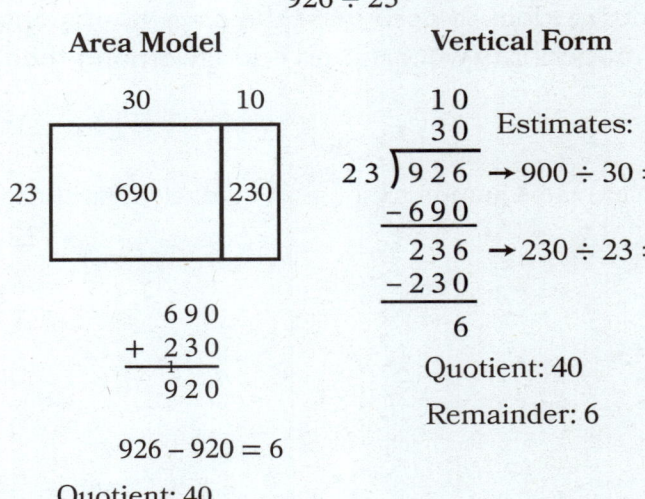

Area Model

Vertical Form

Quotient: 40

Remainder: 6

The total, or number being divided, in a division problem is the dividend. The number of groups or the size of the group is the divisor. The answer to the division problem is the quotient.

Using an area model next to the vertical form of division helps students understand how the numbers relate to each other. Eventually students will rely on the vertical form as their primary method for solving division problems.

At-Home Activity

Division Situations

Look for opportunities to help your student divide with multi-digit numbers in everyday life. Depending on the situation, they can estimate to find the approximate quotient, use a written method to find the exact quotient, or both. Talk about what the quotients and the remainders mean, and how your student knows an answer makes sense.

- Find packages of paper towels or toilet paper at the store or in an advertisement that have 10 or more rolls in the package. Look for the total amount of square feet on the label. Ask your student to divide the total square feet by the number of rolls in the package to find how many square feet are in each roll.

- Research to find the amount of money people in different careers earn in a year. Ask your student to determine the amount of money each of their classmates would get if that quantity were shared equally among them.

© Great Minds PBC

Name

Date

12

1. Complete the tape diagram. Then complete the vertical form and check your work.

$60 \div 20$

a.

60

| 20 | 20 | 20 |

b.

```
      3
20 ) 6 0
   - 6 0
       0
```

Check: $60 = \underline{\ \ 3\ \ } \times 20$

For $60 \div 20$, I ask myself, "How many groups of 20 are in 60?"

I skip-count by twenties to get to 60.

20, 40, 60

I skip-counted 3 times. I know 3 groups of 20 are in 60.
I draw 3 units of 20 in the tape diagram.

60

| 20 | 20 | 20 |

I write the division in vertical form.

```
      3
20 ) 6 0
   - 6 0
       0
```

I check my work by multiplying 3 and 20.

2. Estimate the quotient. Complete the tape diagram. Then complete the vertical form and check your work.

$85 \div 20 \approx$ ___80___ $\div$ ___20___ $=$ ___4___

a.

	85			
20	20	20	20	5

b.

$$20\overline{)85}$$
$$\underline{-80}$$
$$5$$

quotient: 4 (above)

Quotient: ___4___

Remainder: ___5___

Check: $84 =$ ___4___ $\times 20 +$ ___5___

The **dividend** is the number that is divided by another number. 85 is the dividend because it is the total in the tape diagram, and it is being divided by 20.

I round 85 to 80 because 80 is the multiple of 20 that is closest to 85. I ask myself, "How many groups of 20 are in 80?" I skip-count by twenties to get to 80.

$$20, 40, 60, 80$$

I skip-counted 4 times, so I know 4 groups of 20 are in 80. I draw 4 units of 20 in the tape diagram.

I subtract because I need to know how much is remaining: $85 - 80 = 5$.

There are 5 remaining. I do not have enough to make another group of 20, so I have a remainder of 5.

I check my work by multiplying 4 and 20, and then adding 5.

3. Divide. Then check your work.

$156 \div 50$

$$3$$
$$50\overline{)156}$$
$$\underline{-150}$$
$$6$$

Quotient: ___3___

Remainder: ___6___

Check: $3 \times 50 + 6 = 156$

150 is the multiple of 50 that is closest to 156. I ask myself, "How many groups of 50 are in 150?" There are 3 groups of 50 in 150.

I subtract 150 from 156. There are 6 remaining.

I do not have enough to make another group of 50, so I have a remainder of 6.

I check my work by multiplying 3 and 50, and then adding 6.

REMEMBER

4. Use the rectangle to complete parts (a) and (b).

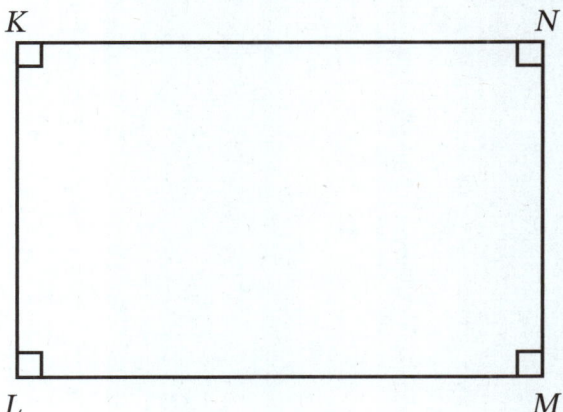

a. Name the pairs of parallel sides shown in the rectangle.

$\overline{KL} \parallel \overline{NM}$, $\overline{KN} \parallel \overline{LM}$

b. Name the pairs of perpendicular sides shown in the rectangle.

$\overline{KL} \perp \overline{KN}$, $\overline{KL} \perp \overline{LM}$, $\overline{NM} \perp \overline{KN}$, and $\overline{NM} \perp \overline{LM}$

Parallel lines never intersect, or cross. They are always the same distance apart, no matter where I measure. $\overline{KL} \parallel \overline{NM}$ means line segment KL is parallel to line segment NM.

Perpendicular line segments intersect and form a right angle. $\overline{KL} \perp \overline{KN}$ means line segment KL is perpendicular to line segment KN.

Name _____   Date _____

1. Complete the tape diagram. Then complete the vertical form and check your work.

   90 ÷ 30

   a. 90
   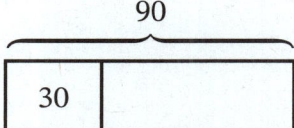

   b.
   $$30\overline{)90}$$

   Check: 90 = _____ × 30

2. Estimate the quotient. Complete the tape diagram. Then complete the vertical form and check your work.

   61 ÷ 30 ≈ _____ ÷ _____ = _____

   a. 61
   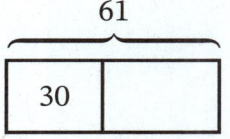

   b.
   $$30\overline{)61}$$

   Quotient: _____

   Remainder: _____

   Check: 61 = _____ × 30 + _____

Divide. Then check your work.

3. 243 ÷ 60

4. 356 ÷ 50

Quotient: _____

Remainder: _____

Check:

Quotient: _____

Remainder: _____

Check:

5. Use the trapezoid to complete parts (a) and (b).

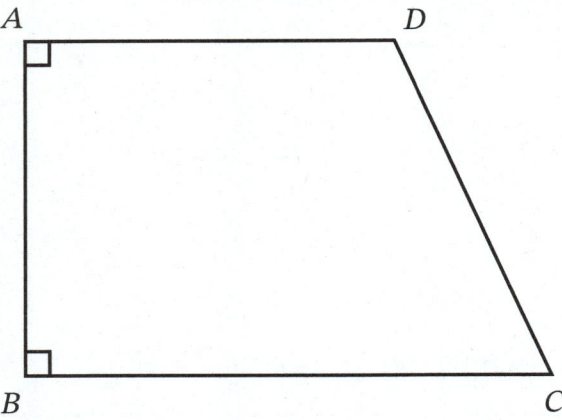

a. Name the pairs of parallel sides shown in the trapezoid.

b. Name the pairs of perpendicular sides shown in the trapezoid.

Name _____ **Date** _____

Estimate the quotient. Complete the tape diagram. Then complete the vertical form and check your work.

1. $99 \div 33 \approx$ ___90___ $\div$ ___30___ $=$ ___3___

a.
```
        99
 ┌──────────────┐
 ┌─────┬─────┬─────┐
 │ 33  │ 33  │ 33  │
 └─────┴─────┴─────┘
```

b.

				3
3	3	9	9
	−	9	9
			0

Check: $99 =$ ___3___ $\times 33$

I round the divisor to 30. I know 90 is a multiple of 30 that is close to 99.
I find $90 \div 30$ and estimate the quotient as 3.
3 groups of 33 is 99. There is no remainder.
I multiply 3 and 33 to check my answer.

2. $41 \div 18 \approx$ ___40___ $\div$ ___20___ $=$ ___2___

a.
```
        41
 ┌──────────────┐
 ┌─────┬─────┬──┐
 │ 18  │ 18  │ 5│
 └─────┴─────┴──┘
```

b.

				2
1	8	4	1
	−	3	6
			5

Quotient: ___2___

Remainder: ___5___

Check: $41 =$ ___2___ $\times 18 +$ ___5___

I round the divisor to 20. I know 40 is a multiple of 20 that is close to 41.
I find $40 \div 20$ and estimate that the quotient is 2.
$2 \times 18 = 36$, so I subtract 36 from 41 to get 5 because I need to know how much remains.
I do not have enough to make another group of 18, so I have a remainder of 5.
I multiply 2 and 18, and then add 5 to check my answer.

3. Divide. Then check your work.

69 ÷ 22

$$22\overline{)69}$$
$$\;\;3$$
$$-66$$
$$\;\;\;3$$

Quotient: ___3___

Remainder: ___3___

Check: $3 \times 22 + 3 = 69$

I estimate the quotient before setting up the vertical form.

I round the divisor to 20 and the dividend to 60.

I know $60 \div 20 = 3$, so I estimate that the quotient is 3.

$3 \times 22 = 66$, so I subtract 66 from 69.

I have a remainder of 3. I do not have enough to make another group of 22.

I multiply 3 and 22, and then add 3 to check my answer.

REMEMBER

4. Add.

$$\frac{8}{10} + \frac{11}{100}$$

$$\frac{91}{100}$$

I know the units must be the same before I add fractions.

I rename tenths as hundredths.

I write an equation by using multiplication. I multiply the numerator and the denominator of $\frac{8}{10}$ by 10 to rename tenths as hundredths.

$$\frac{8}{10} = \frac{10 \times 8}{10 \times 10} = \frac{80}{100}$$

Now I add the hundredths. I know 80 hundredths + 11 hundredths = 91 hundredths.

$$\frac{80}{100} + \frac{11}{100} = \frac{91}{100}$$

5. Use polygons A–F to complete parts (a) and (b).

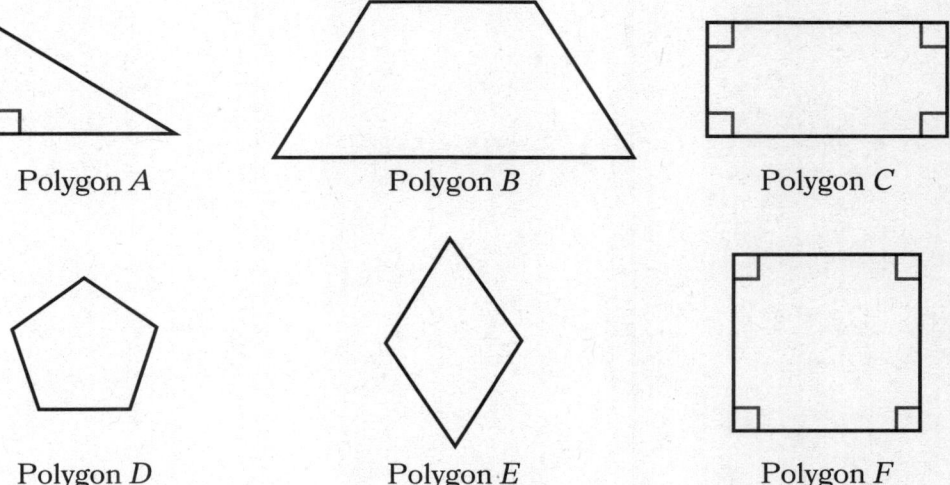

Polygon A   Polygon B   Polygon C

Polygon D   Polygon E   Polygon F

a. Mark the right angles on each polygon.

b. Write the name of each polygon in the category that best describes it.

At Least 1 Pair of Perpendicular Sides	No Pairs of Perpendicular Sides
Polygon A	Polygon B
Polygon C	Polygon D
Polygon F	Polygon E

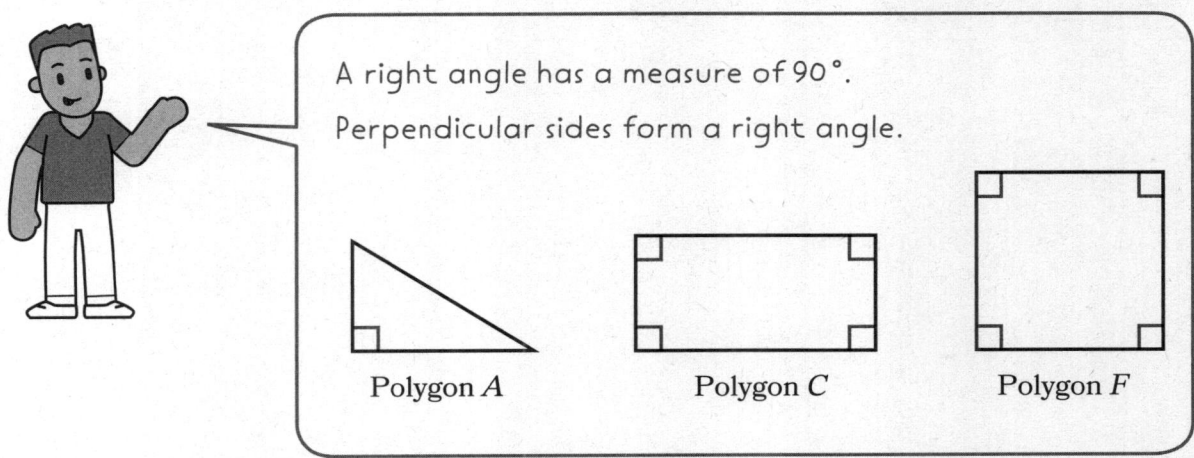

A right angle has a measure of 90°.

Perpendicular sides form a right angle.

Polygon A   Polygon C   Polygon F

Name _____ **Date** _____

Estimate the quotient. Complete the tape diagram. Then complete the vertical form and check your work.

1. $86 \div 43 \approx$ _____ $\div$ _____ $=$ __2__

   a.

   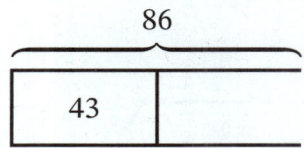

   b.

4	3	)	8	6

   $-$

   Check: $86 =$ _____ $\times 43$

2. $63 \div 19 \approx$ _____ $\div$ _____ $=$ _____

   a.

   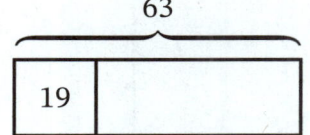

   b.

   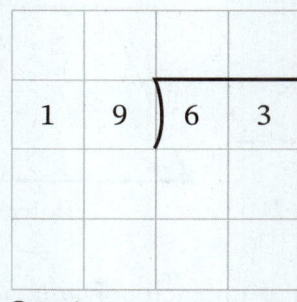

   Quotient: _____

   Remainder: _____

   Check: $63 =$ _____ $\times 19 +$ _____

Divide. Then check your work.

3. $86 \div 21$

   Quotient: _____

   Remainder: _____

   Check:

4. $66 \div 31$

   Quotient: _____

   Remainder: _____

   Check:

REMEMBER

5. Add.

   $\frac{7}{10} + \frac{24}{100}$

6. Use polygons *A–F* to complete parts (a) and (b).

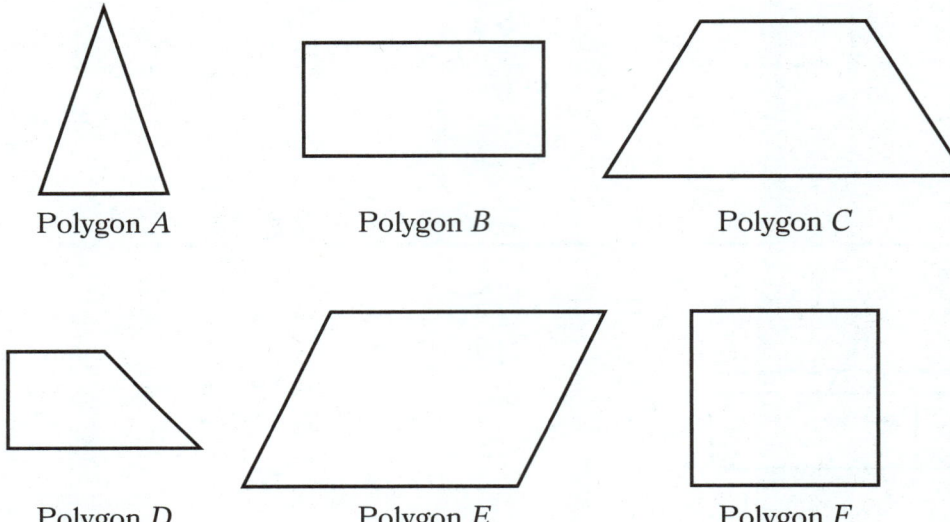

Polygon *A* Polygon *B* Polygon *C*

Polygon *D* Polygon *E* Polygon *F*

   a. Mark the right angles on each polygon.

   b. Write the name of each polygon in the category that best describes it.

At Least 1 Pair of Perpendicular Sides	No Pairs of Perpendicular Sides

Name _____   **Date** _____

1. Estimate the quotient. Then complete the vertical form and check your work. Draw a tape diagram if it helps you divide.

$145 \div 29 \approx$ __150__ $\div$ __30__ $=$ __5__

```
        5
  29) 1 4 5
     -1 4 5
          0
```

Check: $145 =$ __5__ $\times 29$

> I round the divisor to 30. I round 145 to 150 because it is the closest multiple of 30.
>
> I know $15 \div 3 = 5$, so $150 \div 30 = 5$.
>
> I use the estimated quotient, 5, when writing the division in vertical form.
>
> I multiply 5 and 29.
>
> $$5 \times 29 = 145$$
>
> $145 - 145 = 0$, so there is not a remainder.

2. Divide. Then check your work.

$264 \div 42$

```
        6
  42) 2 6 4
     -2 5 2
        1 2
```

Quotient: __2__

Remainder: __12__

Check: $6 \times 42 + 12 = 264$

> I round the divisor to 40. I can round 264 to 240 or 280 because they are the closest multiples of 40. I choose 240 and underestimate.
>
> I know $24 \div 4 = 6$, so $240 \div 40 = 6$.
>
> I use the estimated quotient, 6, when writing the division in vertical form.
>
> I multiply 6 and 42.
>
> $$6 \times 42 = 252$$
>
> $264 - 252 = 12$, so the remainder is 12.

Use the Read–Draw–Write process to solve the problem.

3. Tara has 284 seashells. She can fit 63 seashells in a bucket. She wants to put all her seashells in buckets. What is the fewest number of buckets Tara needs?

284 ÷ 63

Quotient: 4

Remainder: 32

Tara needs 5 buckets.

I read the problem. I read again.

As I reread, I think about what I can draw.

I draw a tape diagram. I label the whole as 284 to represent the total number of seashells. I draw one part and label it 63 to represent the number of seashells in each bucket. I do not know the number of buckets.

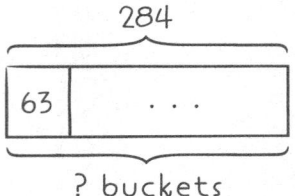

? buckets

I divide 284 by 63. First, I estimate the quotient. I round the divisor to 60. I round 284 to 300 because it is the closest multiple of 60.

300 ÷ 60 = 5, so I multiply 5 by 63 and get 315.

I see that I overestimated, because 315 is more than the dividend, 284.

I try 4 as my estimated quotient instead. I multiply 4 by 63 and get 252.

252 is less than the dividend. I subtract 252 from 284 to get 32. I do not have enough to make another group of 63, so 32 is the remainder.

I know 4 buckets hold 252 shells. Tara needs another bucket to hold the extra 32 shells that are left over. She will need 5 buckets to hold all the shells.

REMEMBER

4. For each figure, draw all lines of symmetry. Circle any figure that does not have a line of symmetry.

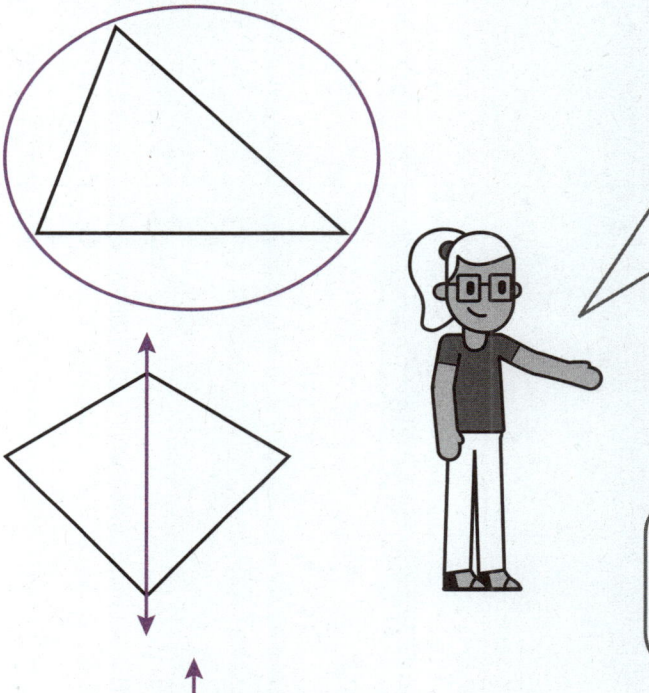

I imagine folding each figure in half. I want to make two parts match exactly on either side of the fold. I can fold horizontally, vertically, or diagonally.

Each fold where the two parts match exactly is a line of symmetry.

A figure can have no lines of symmetry, one line of symmetry, or multiple lines of symmetry.

Name

Date

Estimate the quotient. Then complete the vertical form and check your work. Draw a tape diagram if it helps you divide.

1. $108 \div 18 \approx$ _____ $\div$ _____ $=$ _____

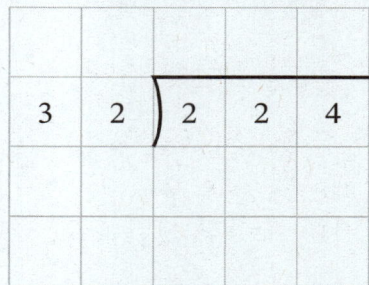

Check: $108 =$ _____ $\times 18$

2. $224 \div 32 \approx$ _____ $\div$ _____ $=$ _____

Check: $224 =$ _____ $\times 32$

Divide. Then check your work.

3. $275 \div 36$

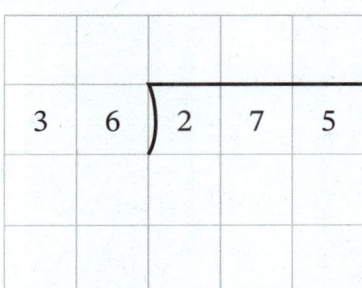

Quotient: _____

Remainder: _____

Check:

4. $387 \div 48$

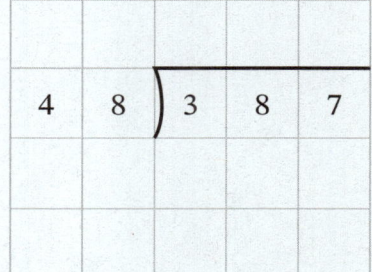

Quotient: _____

Remainder: _____

Check:

Use the Read–Draw–Write process to solve the problem.

5. Ryan has 228 photos. He can fit 42 photos in a photo album. He wants to place all of his photos in albums. What is the fewest number of photo albums Ryan needs?

REMEMBER

6. For each figure, draw all lines of symmetry. Circle any figure that does not have a line of symmetry.

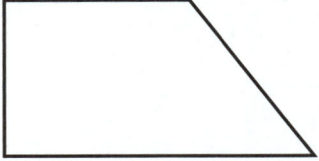

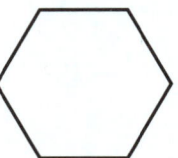

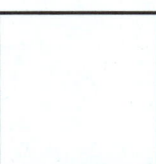

Name

Date

1. Divide by using an area model. Then check your work.

240 ÷ 16

Sample:

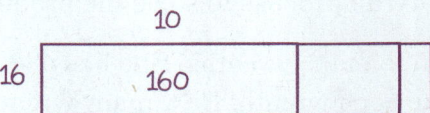

Check:

240 = ___15___ × 16

There are at least 10 groups of 16 in 240 because 10 × 16 = 160. I use 10 as a partial quotient. I write 160 in the first part of the area model.

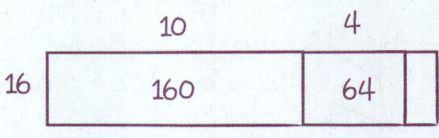

240 − 160 = 80

Now I have 80 left to divide by 16. I can make 4 groups of 16 without going over 80 because 4 × 16 = 64.

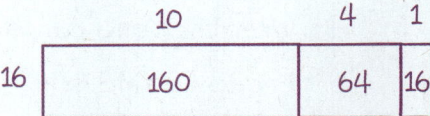

80 − 64 = 16

Now I have 16 left to divide by 16, so I can make 1 group of 16.

10   4 1
16 | 160 | 64 |16|

I add the partial quotients along the top of the area model:

10 + 4 + 1 = 15.

The quotient of 240 divided by 16 is 15.

2. Divide. Then check your work.

$528 \div 22$

					4
				2	0
2	2	)	5	2	8
	−		4	4	0
				8	8
	−			8	8
					0

Check:

528 = ___24___ × 22

I estimate the divisor. I round 22 to 20.

I use the estimate 20 × 20 = 400 to start. I write 20 as a partial quotient. I multiply 20 by 22 to get 440.

I subtract 440 from 528 and 88 is left. I can make 4 groups of 22. I multiply 4 by 22 to get 88. The remainder is 0.

Use the Read–Draw–Write process to solve the problem.

3. Yuna uses ribbon to make wreaths. She has 480 pieces of ribbon. She must use 32 pieces of ribbon to make each wreath. How many wreaths can Yuna make?

$480 \div 32$

Q: 15

R: 0

Yuna can make 15 wreaths.

I read the problem. I read again.

As I reread, I think about what I can draw.
I draw a tape diagram.

I label the whole as 480 to represent the total number of pieces of ribbon. I label one part as 32 to represent how many pieces of ribbon she uses for each wreath. I do not know how many wreaths Yuna can make.

I need to find 480 ÷ 32. I use vertical form to record the division.

I find the quotient by adding the partial quotients: 10 + 5 = 15. There is no remainder.

480

32	. . .

? wreaths

					5
				1	0
3	2	)	4	8	0
	−		3	2	0
			1	6	0
	−		1	6	0
					0

REMEMBER

Use the Read–Draw–Write process to solve the problem. Write the solution statement by using a decimal number.

4. Lacy swims two legs of a race. She swims the first leg in 1.4 minutes and the second leg in 1.68 minutes. How many minutes does Lacy swim altogether during the race?

$$1\frac{4}{10} + 1\frac{68}{100} = 3\frac{8}{100}$$

Lacy swims for 3.08 minutes.

I read the problem. I read again.

As I reread, I think about what I can draw.

I draw a tape diagram. In each part, I show the number of minutes it takes Lacy to swim each leg. I do not know the total, so I label it t.

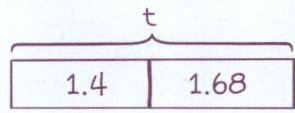

I need to find 1.4 + 1.68. I know 1.4 is 1 and 4 tenths, which is $1\frac{4}{10}$, and 1.68 is 1 and 68 hundredths, or $1\frac{68}{100}$. I rename $\frac{4}{10}$ as $\frac{40}{100}$ so I have like units to add.

$$1\frac{4}{10} + 1\frac{68}{100} = 1\frac{40}{100} + 1\frac{68}{100} = 2 + \frac{108}{100}$$

I know $\frac{100}{100}$ is equal to 1. I know $\frac{108}{100}$ is greater than 1.

I rename $\frac{108}{100}$ as $\frac{100}{100} + \frac{8}{100}$.

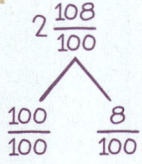

I add the whole numbers and the remaining fraction:
$2 + 1 + \frac{8}{100} = 3\frac{8}{100}$.

I write $3\frac{8}{100}$ as a decimal number. I know $3\frac{8}{100}$ is 3 and 8 hundredths, which is 3.08 as a decimal number.

Name _____ **Date** _____

1. Divide by using an area model. Then check your work.

   $168 \div 12$

   Check:

   $168 = \underline{\qquad} \times 12$

2. Divide. Then check your work.

   $414 \div 18$

   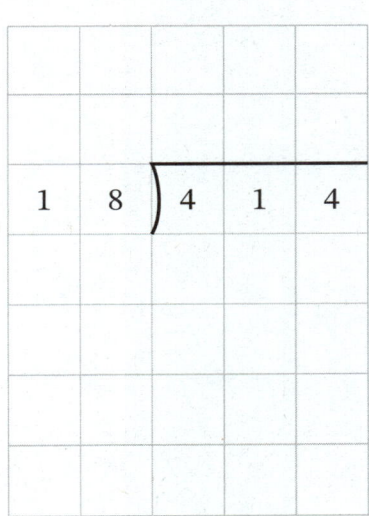

   Check:

   $414 = \underline{\qquad} \times 18$

Use the Read–Draw–Write process to solve the problem.

3. Adesh is making necklaces. He has 336 beads. He must use 28 beads to make each necklace. How many necklaces can Adesh make?

REMEMBER

Use the Read–Draw–Write process to solve the problem. Write the solution statement by using a decimal number.

4. Noah puts 3.57 grams of blueberries in his oatmeal. He feeds his bird 1.3 grams of blueberries. How many grams of blueberries does Noah use altogether?

Name _____ Date _____

16

1. Estimate the partial quotients as you divide. The first estimate is started for you. Make as many estimates as needed. Then check your work.

    $9{,}499 \div 45$

					1
				1	0
			2	0	0
4	5 ⟌	9,	4	9	9
	−	9	0	0	0
			4	9	9
	−		4	5	0
				4	9
	−			4	5
					4

Quotient: ___211___

Remainder: ___4___

Check:

$9{,}499 = $ ___211___ $\times 45 + $ ___4___

Estimates:

→ $8{,}000 \div 40 = \underline{200}$

→ $400 \div 40 = 10$

I estimate: $8{,}000 \div 40 = 200$. I write the partial quotient 200, lining up the place values in vertical form.

I multiply and then subtract to find the amount left over, 499.

I estimate: $400 \div 40 = 10$. I write the partial quotient 10, lining up the place values in vertical form.

I multiply and then subtract to find the amount left over, 49.

I do not need to estimate because I know there is only 1 group of 45 in 49.

I multiply and then subtract to find the amount left over, 4.

I add the partial quotients.

$$200 + 10 + 1 = 211$$

So the quotient is 211 and the remainder is 4.

2. Divide. Then check your work.

1,389 ÷ 43

Check:

$$1{,}389 = 32 \times 43 + 13$$

Quotient: ____32____   Remainder: ____13____

I estimate: 1,200 ÷ 40 = 30. I use the estimated quotient to divide in vertical form. I write the partial quotient 30, lining up the place values. I multiply and then subtract to find the amount left over, 99.

I estimate: 80 ÷ 40 = 2. I write the partial quotient 2, lining up the place values. I multiply and then subtract to find the amount left over, 13.

I do not have enough to make another group of 43, so the remainder is 13.

```
        2
       30
  43 ) 1389
     - 1290
        99
     -  86
        13
```

REMEMBER

3. Compare the fractions by using >, =, or <. Explain your thinking by using pictures, numbers, or words.

$$\frac{1}{2} \ \underline{\ <\ } \ \frac{4}{5}$$

$$\frac{1}{2} = \frac{5 \times 1}{5 \times 2} = \frac{5}{10}$$

$$\frac{4}{5} = \frac{2 \times 4}{2 \times 5} = \frac{8}{10}$$

$$\frac{5}{10} < \frac{8}{10}, \text{ so } \frac{1}{2} < \frac{4}{5}.$$

I find a common denominator to compare fractions.

I know 10 is a multiple of 2 and 5. I rename both fractions as tenths. To compare the fractions, I look at the number of tenths.

I know $\frac{5}{10}$ is less than $\frac{8}{10}$ because 5 is less than 8.

Name _____   Date _____

1. Estimate the partial quotients as you divide. The first estimate is started for you. Make as many estimates as needed. Then check your work.

   $6{,}743 \div 21$

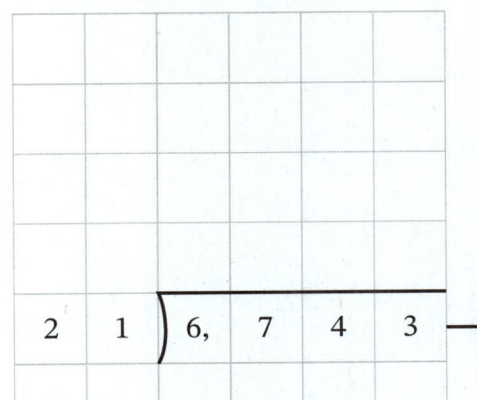

   Check:

   $6{,}743 =$ _____ $\times\ 21 +$ _____

   Estimates:

   $6{,}000 \div 20 =$ _____

   Quotient: _____   Remainder: _____

2. Divide. Then check your work.

   $1{,}687 \div 33$

   Check:

   Quotient: _____   Remainder: _____

REMEMBER

3. Compare the fractions by using >, =, or <. Explain your thinking by using pictures, numbers, or words.

$$\frac{3}{4} \underline{\hspace{2cm}} \frac{2}{6}$$

FAMILY MATH
Multi-Step Problems with Whole Numbers

Dear Family,

Your student is learning how the operations addition, subtraction, multiplication, and division are used in real-world situations. They draw and analyze tape diagrams to see how statements and expressions are related. Then they write word problems to match an expression or tape diagram. They explore how placing parentheses in an expression can change its value. Your student uses operations with whole numbers to solve multi-step word problems. They see there are multiple ways to draw a model to represent a problem.

---

3 times the sum of 15 and 25

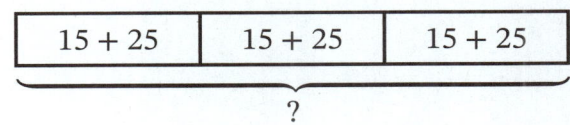

$3 \times (15 + 25)$

The statement, tape diagram, and expression all represent the same mathematical relationship. The parentheses in the expression correspond to the groups shown in the tape diagram.

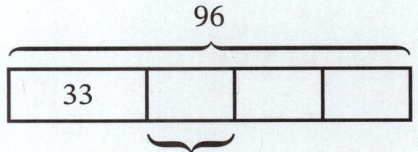

Sample:

Blake makes 96 muffins for the bake sale. He sells 33 of them and puts the remaining muffins in 3 containers to take home. If he puts the same number of muffins in each container, how many muffins are in each?

The sample shows a word problem that can be represented by this tape diagram. As long as the relationship between the numbers stays the same, the context of the problem can change.

At-Home Activity

Birthday Word Problem

Have your student write their birthday or another important date in number form by using four digits for the year. Ask them to use the numbers in the date to create an expression that includes two different operations (+, −, ×, or ÷) and parentheses.

- For the date August 19, 2011, your student would write an expression with the numbers 8, 19, and 2,011 such as $2,011 − (8 \times 19)$.

Then have them write a word problem that matches their expression.

- A word problem to match $2{,}011 - (8 \times 19)$ is: I scored 2,011 points in my video game. Then I got to the last level. I tried to pass the level and failed 8 times. Each time I tried to pass the level and failed, I lost 19 points. How many points do I have now?

Finally, they can evaluate the expression and solve their word problem.

- $$2{,}011 - (8 \times 19) = 2{,}011 - 152$$
  $$= 1{,}859$$

  I have 1,859 points now.

© Great Minds PBC

Name _____   Date _____

1. Draw a tape diagram and write an expression to represent the statement.

    The sum of three 12s and four 8s

12	12	12	8	8	8	8

    $(3 \times 12) + (4 \times 8)$

    I draw a tape diagram with 3 units of 12 and 4 units of 8.

    I represent the 3 units of 12 as 3 × 12.

    I represent the 4 units of 8 as 4 × 8.

    I write an addition expression and use parentheses to show how to group the numbers.

2. Write a statement and an expression to represent the tape diagram. Then evaluate your expression.

41	19	41	19	41	19

    Statement: 3 times as much as the sum of 41 and 19

    Expression: $(41 + 19) \times 3$

    Value of Expression: 180

    The tape diagram shows 3 times as much as the sum of 41 and 19.

    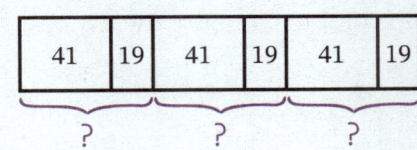

    I write (41 + 19) × 3 to represent the tape diagram. I write parentheses around 41 + 19 to show that I need to find the sum before I multiply by 3.

    To evaluate the expression, or find its value, first I find 41 + 19. Then I multiply the sum by 3.

    $(41 + 19) \times 3 = 60 \times 3 = 180$

3. Place parentheses to make the equation true.

$$12 \times 2 + 5 + 4 = 88$$

$12 \times (2 + 5) + 4 = 88$

I model the equation with tape diagrams to determine where the parentheses should go.

When I group 12 × 2 with parentheses, the tape diagram shows 2 groups of 12, a group of 5, and a group of 4. The total is 33 and not 88 like in the equation. The parentheses should not go around 12 × 2.

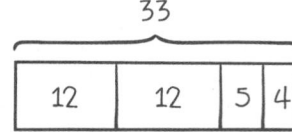

When I group 2 + 5 with parentheses, the tape diagram shows 12 groups of 2 + 5 and a group of 4. The total is 88, like in the equation. The parentheses should go around 2 + 5.

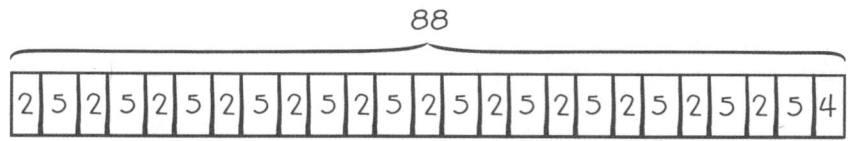

4. Use >, =, or < to compare the expressions. Explain how you can compare the expressions without evaluating them.

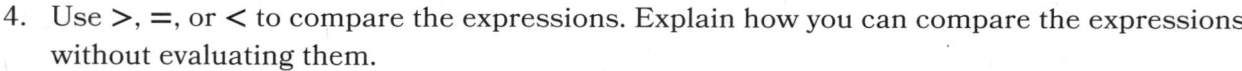

$$(40 \times 4) + (12 \times 4) \; \boxed{>} \; (30 \times 4) + (5 \times 4)$$

I can think about groups of 4 to compare the expressions. 52 fours is greater than 35 fours.

I see 4 is a factor in each group in the expressions.

I think about the groups as 40 fours, 12 fours, 30 fours, and 5 fours.

40 fours + 12 fours = 52 fours

30 fours + 5 fours = 35 fours

52 fours is greater than 35 fours.

REMEMBER

5. Consider the number shown.

    $\boxed{6}$63,849

    a. Draw a box around the digit that represents 10 times as much as the underlined digit.

    b. Complete the equations to show the relationship between the boxed and underlined digits.

    $$\underline{600,000} = 10 \times \underline{60,000}$$

    $$\underline{600,000} \div 10 = \underline{60,000}$$

Each 6 represents a different value.

The underlined 6 is in the ten-thousands place. It represents 60,000.

I multiply to show that 600,000 is 10 times as much as 60,000.

I draw a box around the 6 in the hundred-thousands place.

I know 10 × 60,000 = 600,000.

So, 600,000 ÷ 10 = 60,000.

Name _____   Date _____

Draw a tape diagram and write an expression to represent each statement.

1. The sum of three 16s and five 13s

2. 5 times as much as the sum of 14 and 6

3. Write a statement and an expression to represent the tape diagram. Then evaluate your expression.

| 22 | 9 | 22 | 9 | 22 | 9 | 22 | 9 | 22 | 9 | 22 | 9 |

Statement: _____

_____

Expression: _____

Value of Expression: _____

4. Place parentheses to make the equation true.

$$11 \times 6 + 2 + 2 = 90$$

5. Use >, =, or < to compare the expressions. Explain how you can compare the expressions without evaluating them.

$$(42 \times 8) - (17 \times 8) \bigcirc (14 \times 8) + (15 \times 8)$$

REMEMBER

6. Consider the number shown.

   2<u>2</u>9,576

   a. Draw a box around the digit that represents 10 times as much as the underlined digit.

   b. Complete the equations to show the relationship between the boxed and underlined digits.

$$\underline{\hspace{2cm}} = 10 \times \underline{\hspace{2cm}}$$

$$\underline{\hspace{2cm}} \div 10 = \underline{\hspace{2cm}}$$

18

Name _____ Date _____

1. Write an expression that represents the tape diagram. Then write a word problem that can be represented by the tape diagram and expression.

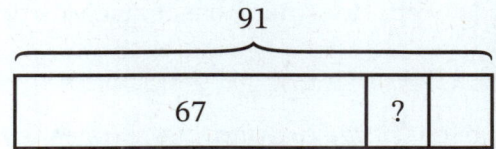

$(91 - 67) \div 2$

Sample: Lisa has 91 eggs. She sells 67 eggs at a farmers market and divides the remaining eggs equally into 2 containers. How many eggs does Lisa put in each container?

The total is 91.

One part is 67. The other part is divided into 2 equal parts.

I use the tape diagram to write an expression.

I put parentheses around the subtraction expression to show that the subtraction is done before the division.

The word problem I write begins with someone having 91 objects. First, 67 objects are taken away. Then the remaining objects are divided into 2 equal groups.

2. Consider the expression.

$$(26 - 19) + (3 \times 13)$$

Write a word problem that can be represented by the given expression.

Sample: Leo has 26 baseball cards. He gives 19 baseball cards to his friend. Then he buys 3 packs of 13 baseball cards each. How many baseball cards does Leo have now?

I see parentheses around the expression 26 – 19, which tells me it is a unit.

The expression 3 × 13 is also a unit. I know it is a unit because it is grouped with parentheses.

The word problem I write begins with 26 objects and 19 of the objects in the group are taken away. There are also 3 groups of 13 of the same objects. The total number of objects in my word problem is the sum of the values of each group.

REMEMBER

3. Consider the expression shown.

$$100 \times 10^5$$

How does the exponent help you think about shifting the digits in the first factor to find the product?

The exponent of 5 helps me know I need to shift the digits 5 places. Because I am multiplying 100 by 10 five times, I need to shift the digits 5 places to the left.

The exponent tells me how many times to use 10 as a factor.

$10^5 = 10 \times 10 \times 10 \times 10 \times 10$

When multiplying by a power of 10, I know I shift the digits the same number of places as the exponent. I shift the digits to the left when I multiply.

$$100 \times 10^5 = 10,000,000$$

4. Use $\overleftrightarrow{GH}$ to complete parts (a) and (b).

Sample:

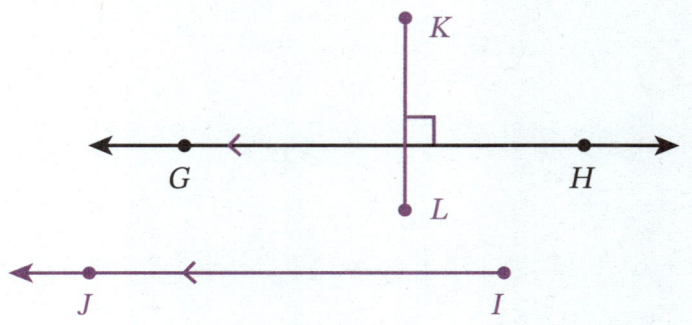

a. Draw $\overrightarrow{IJ}$ parallel to $\overleftrightarrow{GH}$.

b. Draw $\overline{KL}$ perpendicular to $\overleftrightarrow{GH}$.

I use a right-angle tool and a straightedge to draw a parallel ray. Parallel lines never intersect.

I line up my right-angle tool along $\overleftrightarrow{GH}$ so the top edge touches the line. Then I hold my straight edge along the other edge of the right-angle tool.

I slide my right-angle tool down along the straightedge. I trace along the top edge to draw $\overrightarrow{IJ}$. I start the ray at point I. I extend the ray through point J with an arrowhead at the end. I mark the lines to show they are parallel.

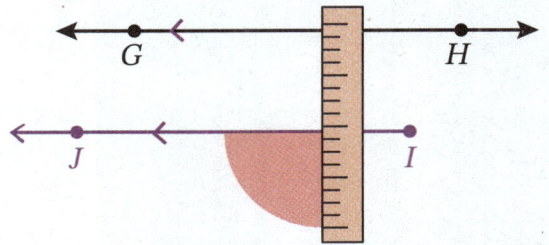

I use a right-angle tool and a straightedge to draw a perpendicular line segment. The line and line segment intersect at right angles.

I line up one edge of the right-angle tool along $\overleftrightarrow{GH}$.

I trace along the side of the right-angle tool that is perpendicular to $\overleftrightarrow{GH}$.

I label the ends of the line segment with point K and point L. I mark the angle to show the lines are perpendicular.

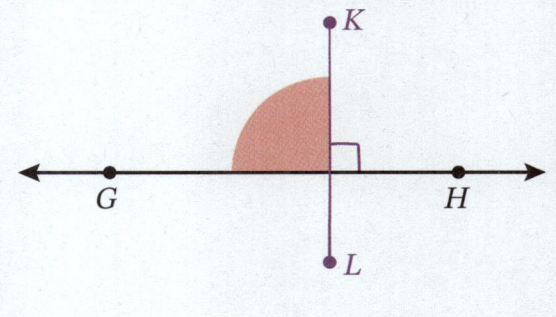

18

Name _____ Date _____

Write an expression that represents each tape diagram. Then write a word problem that can be represented by the tape diagram and expression.

1.

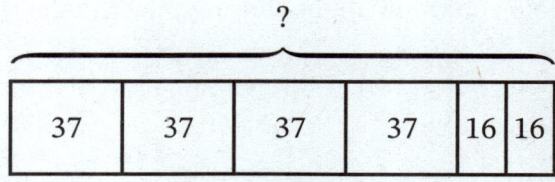

2.

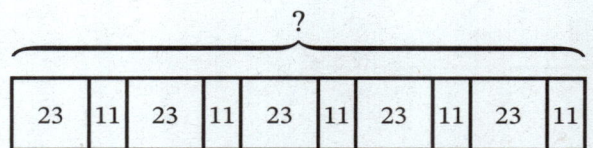

Consider each expression. Write a word problem that can be represented by the given expression.

3. $(18 - 6) \div 4$

4. $(14 - 12) \times 15 + 20$

REMEMBER

5. Consider the expression shown.

$$10{,}000 \times 10^4$$

How does the exponent help you think about shifting the digits in the first factor to find the product?

6. Use $\overleftrightarrow{ST}$ to complete parts (a) and (b).

$$S \qquad\qquad\qquad\qquad T$$

a. Draw $\overleftrightarrow{VW}$ parallel to $\overleftrightarrow{ST}$.

b. Draw $\overrightarrow{TU}$ perpendicular to $\overleftrightarrow{ST}$.

Name _____   Date _____

Use the Read–Draw–Write process to solve each problem.

1. Eddie has 13 tomato plants. There are 24 tomatoes on each plant. He puts an equal number of tomatoes in 26 containers to give to family and friends. If Eddie wants to give away all the tomatoes, how many should he put in each container?

$13 \times 24 = 312$

$312 \div 26 = 12$

Eddie should put 12 tomatoes in each container.

> I read the problem. I read again.
>
> As I reread, I think about what I can draw.
>
> I draw two tape diagrams to represent the problem.
>
> First, I draw a tape diagram to represent that each of the 13 tomato plants has 24 tomatoes on it. The total number of tomatoes is unknown.

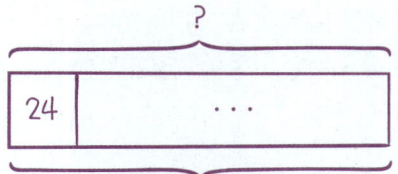

I multiply to find the total number of tomatoes. I use an area model to find the product.

```
      20    4
   ┌─────┬─────┐     72
 3 │  60 │  12 │   + 240
   ├─────┼─────┤   ───────
10 │ 200 │  40 │     312
   └─────┴─────┘
```

The total number of tomatoes is 312.

> Then I draw another tape diagram to represent the 312 tomatoes divided into equal-size groups. There are 26 groups. The size of each group is unknown.

312 tomatoes

```
┌───┬───────────┐
│ ? │    ...     │
└───┴───────────┘
```

26 containers

I divide to find the number of tomatoes in each container. I divide by finding partial quotients.

2. There are 2,475 people sitting in a theater. An equal number of people sit in each of the theater's
    11 sections. The tickets for a seat in section C each cost $57. What is the total cost of the tickets
    for the people sitting in section C?

    $2{,}475 \div 11 = 225$

    $225 \times 57 = 12{,}825$

    The total cost of the tickets in section C is $12,825.

I read the problem. I read again.

As I reread, I think about what I can draw. I draw two tape diagrams to represent the problem.

I draw a tape diagram to represent 2,475 people divided into 11 sections. The number of people in each section is unknown. Section C is one of the sections, so one part represents the people sitting in section C.

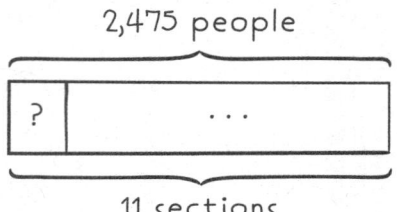

2,475 people

? . . .

11 sections

I divide to find the number of people sitting in one section.

Then I draw a second tape diagram to represent 225 people with tickets that each cost $57. The total cost is unknown.

?

57 . . .

225 people

I multiply to find the total cost of the tickets in section C.

REMEMBER

3. Rewrite the expression by using an exponent.

$$9 \times 10 \times 10 \times 10 \times 10 \times 10 = 9 \times \underline{\ 10^5\ }$$

I know that an exponent represents how many times I multiply by the same number. I count the number of times 10 is a factor.

There are 5 tens, so the exponent is 5. I write 10 to represent the number being multiplied. I write 5 as an exponent to represent the number of times 10 is a factor.

4. Use a protractor to measure the angle.

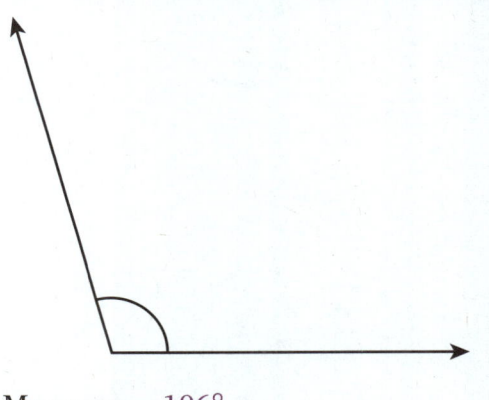

Measure: __106°__

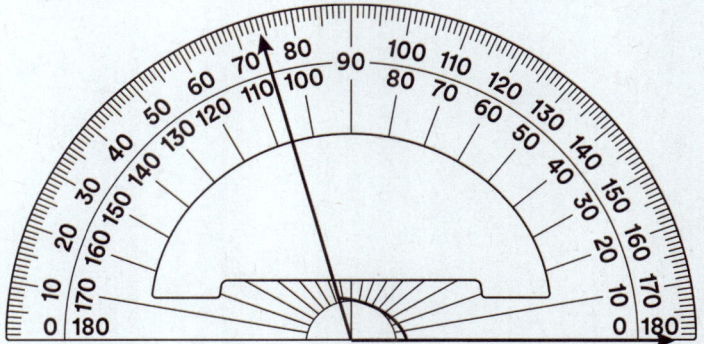

I line up my protractor so one ray goes through 0°. The vertex of the angle must be where the zero line and the 90° line meet.

I look for where the ray crosses the scale. If I need to, I extend the rays with a straightedge so that I can measure the angle.

My angle is obtuse, so I use the scale that shows a measure greater than 90°.

I count one tick mark from 105° to determine that the measure of the angle is 106°.

19

Name _____ Date _____

Use the Read–Draw–Write process to solve each problem.

1. A bookstore has 11 rows of shelves. Each row has 13 shelves. There are 3,146 books in the store. An equal number of books are placed on each shelf. How many books are on each shelf?

2. A factory makes 3,876 bags of rope. Boxes that each hold 19 bags of rope are sold for $15 per box. If the factory sells all the boxes of rope, how much money does the factory earn?

3. Toby buys 16 packs of trading cards. Each pack has 15 cards. He puts all the trading cards in a notebook. He fills 20 pages of it with an equal number of cards on each page. How many trading cards are on each page?

REMEMBER

4. Rewrite the expression by using an exponent.

$$8 \times 10 \times 10 \times 10 \times 10 \times 10 \times 10 \times 10 = 8 \times \underline{\qquad}$$

5. Use a protractor to measure the angle. A paper protractor is included, if needed.

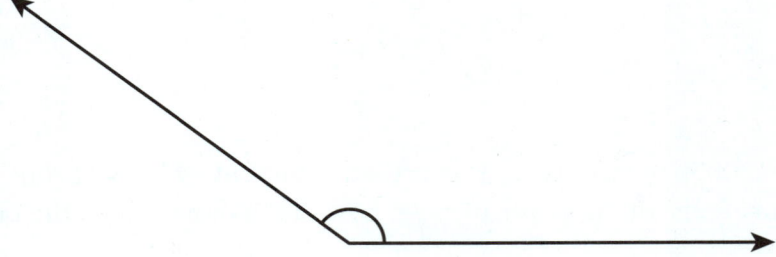

Measure: _____

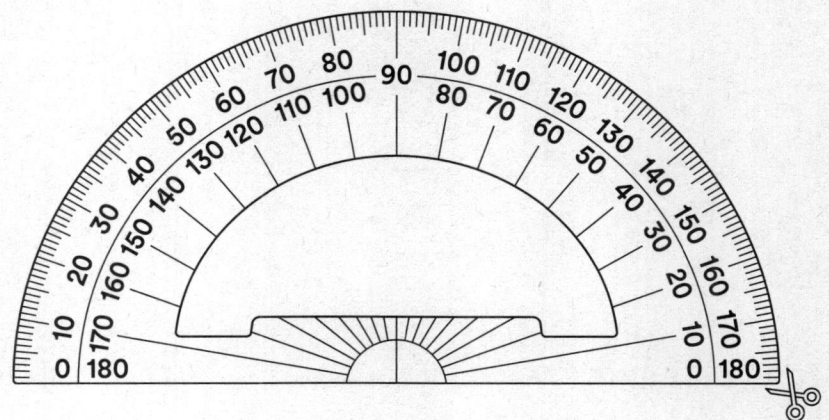

Name _____

Date _____

Use the Read–Draw–Write process to solve each problem.

1. The town of Harper has a population of 8,268 people. The town of Walden has a population of 1,884 fewer people than Harper. The population of Walden is 3 times as large as the population of Franklin. What is the population of Franklin?

$8,268 - 1,884 = 6,384$

$6,384 \div 3 = 2,128$

The population of Franklin is 2,128 people.

I read the problem. I read again.

As I reread, I think about what I can draw.

First, I draw a tape diagram to represent the population of Harper: 8,268. I draw a second tape diagram to represent that Walden has 1,884 fewer people than Harper.

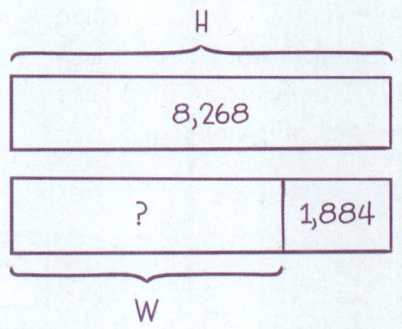

I subtract to find the population of Walden.

Next, I drew a tape diagram to represent the population of Walden: 6,384. I partition the tape into 3 equal parts. 1 part represents the population of Franklin.

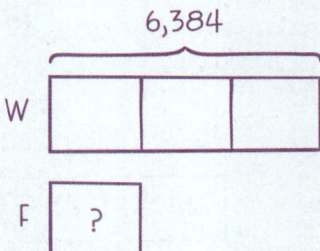

I divide to find the population of Franklin.

2. A parking garage has 728 spaces. There are an equal number of spaces on each of the 4 floors of the garage. The parking garage charges $12 a day to park. If all the spaces are full, how much money can the third floor earn in 1 week?

$728 \div 4 = 182$

$182 \times 12 = 2{,}184$

$2{,}184 \times 7 = 15{,}288$

The third floor can earn $15,288 in 1 week.

I read the problem. I read again.

As I reread, I think about what I can draw.

I draw a tape diagram to represent the number of spaces in the parking garage: 728. I partition the tape diagram into 4 equal parts to represent the floors of the garage. The size of 1 part is unknown.

728

?			

I divide to find the number of spaces on each floor.

Next, I draw a tape diagram to represent the amount of money each floor earns in 1 day. 1 part represents the cost to park in each space: $12. The number of parts is the number of spaces on each floor: 182. The total is unknown.

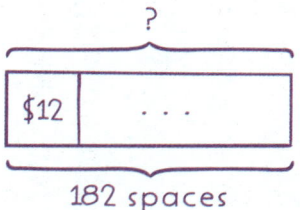

?

$12	. . .

182 spaces

I multiply to find the amount each floor earns in 1 day.

Then I draw a tape diagram to represent the amount of money the third floor can earn in 1 week. 1 part represents the amount the third floor can earn in 1 day: $2,184. There are 7 parts because there are 7 days in 1 week.

?

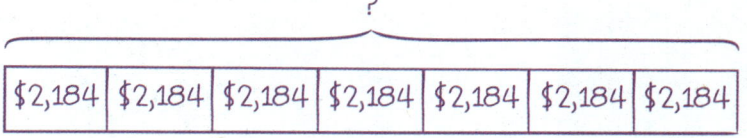

$2,184	$2,184	$2,184	$2,184	$2,184	$2,184	$2,184

I multiply to find the amount of money the third floor can earn in 1 week.

REMEMBER

Use the Read–Draw–Write process to solve the problem.

3. Blake hits a golf ball 23 m 41 cm. On his next try, he hits it 1,560 cm. On his third try, Blake hits the ball twice as far as his first two tries combined. How far does Blake hit the golf ball on his third try?

    23 m 41 cm = 2,341 cm

    2,341 + 1,560 = 3,901 cm

    3,901 × 2 = 7,802 cm

    Blake hits the golf ball 7,802 centimeters on his third try.

I read the problem. I read again.

As I reread, I think about what I can draw.

I draw a tape diagram to represent the distances of Blake's first and second tries combined. The total distance is unknown.

I convert 23 meters 41 centimeters to centimeters so the measurements are named in the same unit.

$$23 \text{ m } 41 \text{ cm} = 23 \text{ m} + 41 \text{ cm}$$
$$= (23 \times 100 \text{ cm}) + 41 \text{ cm}$$
$$= 2,300 \text{ cm} + 41 \text{ cm}$$
$$= 2,341 \text{ cm}$$

2,341	1,560

I add to find the total distance of Blake's first and second tries.

I know that twice as far means 2 times as far. I show 2 groups of 3,901 to represent Blake's third try.

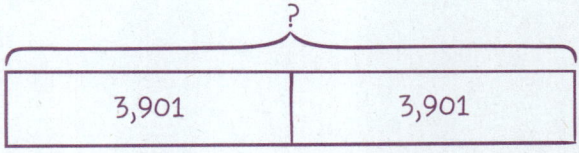

3,901	3,901

I multiply to find the distance of the third try.

PRACTICE PARTNER

20

Name _____   Date _____

Use the Read–Draw–Write process to solve each problem.

1. Tara is on a 6-day road trip. She drives a total of 2,318 miles on the trip. On the first day, she drives 343 miles. She drives the same number of miles on each of the other 5 days. How many miles does Tara drive on each of the other days?

2. The art museum sells a total of 2,790 tickets over Friday, Saturday, and Sunday. The same number of people visit each day. Tickets to the museum cost $8. How much money does the museum make on Saturday and Sunday altogether?

Use the Read–Draw–Write process to solve the problem.

3. A rug is 2 meters 18 centimeters long. Another rug is 124 centimeters long. The hallway is twice the combined length of the rugs. How many centimeters long is the hallway?

Acknowledgments

Kelly Alsup, Adam Baker, Agnes P. Bannigan, Christine Bell, Reshma P Bell, Joseph T. Brennan, Dawn Burns, Amanda H. Carter, David Choukalas, Mary Christensen-Cooper, Nicole Conforti, Cheri DeBusk, Lauren DelFavero, Jill Diniz, Mary Drayer, Christina Ducoing, Karen Eckberg, Melissa Elias, Danielle A Esposito, Janice Fan, Scott Farrar, Gail Fiddyment, Ryan Galloway, Krysta Gibbs, January Gordon, Torrie K. Guzzetta, Kimberly Hager, Jodi Hale, Karen Hall, Eddie Hampton, Andrea Hart, Stefanie Hassan, Tiffany Hill, Christine Hopkinson, Rachel Hylton, Travis Jones, Laura Khalil, Raena King, Jennifer Koepp Neeley, Emily Koesters, Liz Krisher, Leticia Lemus, Marie Libassi-Behr, Courtney Lowe, Sonia Mabry, Bobbe Maier, Ben McCarty, Maureen McNamara Jones, Ashley Meyer, Pat Mohr, Bruce Myers, Marya Myers, Kati O'Neill, Darion Pack, Geoff Patterson, Victoria Peacock, Maximilian Peiler-Burrows, Brian Petras, April Picard, Marlene Pineda, DesLey V. Plaisance, Lora Podgorny, Janae Pritchett, Elizabeth Re, Meri Robie-Craven, Deborah Schluben, Colleen Sheeron-Laurie, Michael Short, Erika Silva, Jessica Sims, Tara Stewart, Heidi Strate, Theresa Streeter, Mary Swanson, James Tanton, Cathy Terwilliger, Saffron VanGalder, Rafael Vélez, Allison Witcraft, Jim Wright, Jackie Wolford, Caroline Yang, Jill Zintsmaster

Trevor Barnes, Brianna Bemel, Lisa Buckley, Adam Cardais, Christina Cooper, Natasha Curtis, Jessica Dahl, Brandon Dawley, Delsena Draper, Sandy Engelman, Tamara Estrada, Soudea Forbes, Jen Forbus, Reba Frederics, Liz Gabbard, Diana Ghazzawi, Lisa Giddens-White, Laurie Gonsoulin, Nathan Hall, Cassie Hart, Marcela Hernandez, Rachel Hirsh, Abbi Hoerst, Libby Howard, Amy Kanjuka, Ashley Kelley, Lisa King, Sarah Kopec, Drew Krepp, Crystal Love, Maya Márquez, Siena Mazero, Cindy Medici, Ivonne Mercado, Sandra Mercado, Brian Methe, Patricia Mickelberry, Mary-Lise Nazaire, Corinne Newbegin, Max Oosterbaan, Tamara Otto, Christine Palmtag, Andy Peterson, Lizette Porras, Karen Rollhauser, Neela Roy, Gina Schenck, Amy Schoon, Aaron Shields, Leigh Sterten, Mary Sudul, Lisa Sweeney, Samuel Weyand, Dave White, Charmaine Whitman, Nicole Williams, Glenda Wisenburn-Burke, Howard Yaffe

Credits

For a complete list of credits, visit http://eurmath.link/media-credits